The Secret Life of Lobsters

The Secret Life of Lobsters

How Fishermen and Scientists Are Unraveling the Mysteries of Our Favorite Crustacean

TREVOR CORSON

HarperCollins*Publishers*

HarperCollins books may be purchased for educational, business, or sales promotional use. For information, please write: Special Markets Department, HarperCollins Publishers Inc., 10 East 53rd Street, New York, NY 10022.

Portions of this book originally appeared in different form in *The Atlantic Monthly* and were reprinted as a selection in *The Best American Science Writing 2003* (published by Ecco, an imprint of HarperCollins Publishers).

FIRST EDITION

Designed by C. Linda Dingler

Printed on acid-free paper

Library of Congress Cataloging-in-Publication Data is available upon request.

ISBN 0-06-055558-0

04 05 06 07 08 ❖/RRD 10 9 8 7 6 5 4

It is in the unexpected or neglected place
that you will find the lobster.

—IRISH SAYING

Contents

PART SIX: *Brooding*

The Secret Life of Lobsters

PROLOGUE

Setting Out, 2001

The morning sky was glowing pink in the southeast but a chill hung in the salt air. The grumble of a truck engine echoed across the harbor. Bruce Fernald's rust-encrusted Ford pickup skidded to a halt in the gravel near the fishermen's co-op on Little Cranberry Island.

Bruce's sternman, Jason Pickering, was waiting on the wharf. Bruce had employed fifteen different sternmen in his thirty years of lobstering, and the most reliable had been the one woman he'd ever hired. After hauling traps with her by his side, through icy gales and summer afternoons suffused with the stench of bait, Bruce had asked Barb to marry him. It was a hard act to follow.

Bruce and Jason rowed across the harbor and clambered aboard Bruce's lobster boat, white with red trim. He'd had her built after the birth of his twin sons, and had christened her the *Double Trouble*. She was fast, though to accomplish her speed she was narrow in the stern, making her tippy when she got sideways to a rough sea. Bruce needed a calm sea this morning, because thirty-three of his eight hundred lobster traps were piled in a pyramid in the stern, along with a couple of miles of coiled rope and unwieldy bundles of buoys.

A century earlier, three hundred Maine islands had been home to year-round communities of fishermen and seafarers. Little Cranberry Island was one of just fourteen such year-round communities that remained. A mile and a half long and shaped like a pork chop, it lay among four other small islands that together formed the Cranberry Isles. Nestled just south of the larger island of Mount Desert, the Cranberries were visible to hikers in Acadia National Park as a cluster of green slabs on the ocean.

Little Cranberry had been Bruce's home for most of his fifty years, and he'd spent most of his adult life trapping lobsters around the island's shores. So had his father, his grandfather, two of his brothers, and the dozen other lobstermen that made their living there. Along with a few builders, artisans, and retirees, and two schoolteachers, the fishermen and their families formed a community so tight that doors were seldom locked. Social life revolved around the general store in the center of the island, where Soos sold groceries and served pizza for lunch. In a corner of the store was the post-office window, where Joy dispensed stamps, local news, and homemade cream puffs. Next door was the two-room schoolhouse where the island's eleven students, from prekindergarten through eighth grade, attended class. Around the corner was the Grange hall, now home to town meetings, potluck dinners, and aerobics classes. Attached to the hall was a small library. Down the main street was the Protestant church. In the other direction was the Catholic chapel, where a fisherman's net hung behind Jesus, the fisher of men. Bruce Fernald attended neither, but if the lobstering didn't improve, it was possible he'd begin attending both.

Bruce plunked his lunch bag on the *Double Trouble*'s forward bulkhead, then yanked off his cap and used it to swat a cloud of mosquitoes in the cabin.

"Ain't they something awful?" he asked Jason.

"Yup," the younger man answered, adjusting the knife on his belt.

Bruce fired up the boat's diesel engine. The *Double Trouble* coughed, cleared a black cloud from her exhaust stack, and thundered to life. The engine's metallic growl ricocheted off the shore and scattered gulls that had been roosting on the bow. Both men clambered into orange rubber overalls, clammy with dew. Jason lifted the lid off the bait bin, filling the cabin with the stench of five-day-old fish.

Bruce propped open a panel of the windshield to draw in fresh air, then nudged the boat into gear. Pulling alongside the thick mooring chain that tethered the boat to a two-ton slab of granite resting on the bottom of the harbor, he freed the vessel and motored away, checking his electronics while waiting for the engine to warm. Mounted on the bulkhead and hanging from the ceiling were a color Fathometer, a depth sounder, a radar unit, a loran navigational locator, a Global Positioning System satellite plotter, and two VHF marine radios. Wires snaked across the interior woodwork, met each other in bundles, and bored up through the ceiling to feed a roof bristling with antennas and down through the hull to supply the underwater transducers. In front of Bruce, between the throttle and the steering wheel, was a white compass the size of a softball. Mounted in a corner was a horn. It would blare if water tripped a switch in the bilge. Generally that would mean the boat was sinking.

As the *Double Trouble* chugged along the island's western shore, Bruce and Jason gazed over the beach toward the dawn, sniffing for wind. Just past the beach, a row of gravestones stood like sentinels over the harbor. Bruce's grandfather lay under a pink square of polished granite, guarded by a field of goldenrod and Queen Anne's lace. Nearby was a lobsterman who'd been battered to death when a storm rolled his boat over the rocks. Under another stone was a sternman who'd drowned three years ago. Other graves revealed that the ocean wasn't the only danger an islander could face. Jason's half sister was buried there. She'd been killed in a car crash on the mainland, just shy of her twentieth birthday.

Rounding a point of land, Bruce piloted the boat into a narrow channel called the Gut. The *Double Trouble*'s sophisticated electronics were no help here. Instead, Bruce peered through the open window and located Asshole Rock, a cracked ledge at the far end of the Gut. Asshole Rock served as a low-tech warning beacon—if a lobsterman saw that the crack was completely exposed at low tide, he knew his boat was likely to run aground, and he would turn around. Bruce saw that the bottom of the crack was still submerged, indicating sufficient draft. He steered forward through the boulders, peering not at the rocks ahead but backward at a church steeple on an island three miles away, which he aligned with a seaweed-covered stone a hundred yards off the boat's stern. A moment later he turned and squinted at a ledge five hundred yards off the bow. When the ledge lined up with the tree line on an island two miles to the east, he turned the boat forty degrees to port.

The *Double Trouble* entered an ocean afire. The sun was emerging from the sea and had stained a bank of clouds to the south yellow, which in turn had gilded the water to the horizon. Seals sprawled on a ledge, and cormorants perched on the rocks, holding their wet wings open to the morning sun like capes.

Bruce gunned the boat to cruising speed for the run offshore, and the growl of the engine widened into a roar. He was leaning over to choose a fresh pair of work gloves when the boat jerked sideways. Bruce had a steady fisherman's physique—stocky, with powerful shoulders—but he was thrown off balance. He swore and yanked back on the throttle. There was no wind, but outside the harbor the sea was a frothing cauldron.

"What's this slop doing out here?" Bruce asked. Perhaps he was addressing the question to his ancestors, who had fished these waters for a hundred years. Jason pursed his lips while Bruce watched the way the sea was working. Packs of wavelets scurried across larger waves at chaotic angles. The *Double Trouble* tossed like a toy.

"There must be a storm rolling this shit in from way offshore," Bruce said, "and coming crossways at the tide." He shot a despairing glance at the pile of gear in the stern. "Why? Oh, why?"

Tightening his grip on the wheel, Bruce nudged the throttle back up. He spread his feet apart and settled into a crouch for his knees to absorb the beating. The *Double Trouble* skidded, bounced, and bucked her way south into the open ocean. It is sometimes said that lobstermen are the cowboys of the American East. The resemblance can be striking.

Facing aft, Jason leaned into the bait bin and stuffed knit bags with fish parts. Normally he would fill the bags throughout the day, but with thirty-three traps to set he needed a head start. Rancid brown juice sloshed over the hems of his gloves and down between his fingers. When a steep wave struck the boat, droplets of bait juice splattered onto his face.

Bruce plucked a tattered notebook from the bulkhead and flipped through pages of scrawled notes, scanning for the coordinates of a particular underwater valley. With a pencil he jotted a few numbers directly onto the white paint of the bulkhead, then squinted up at the GPS plotter above his head. He pressed a few buttons to call up a waypoint, then adjusted his course by several degrees. It was reassuring to see his position confirmed by transmissions from four different satellites, but in a pinch he could go back to navigating by local landmarks and his compass, as his father still did.

Ten minutes later Bruce throttled down and the boat buried her nose in a trough. Jason dunked his gloves into a barrel of steaming water, heated by a coil from the engine, while Bruce stared at blotches of color scrolling across his Fathometer screen. The screen painted the bottom as a jagged black line that marked it as rocky ledge. He circled the boat a quarter turn and motored slowly east, watching the line drop off and the color lighten from black to purple, indicating a deeper section of cobble. As the boat continued east the color changed to orange, indicating gravel. Then the line fell precipitately and settled into a mushy yellow haze, a bottom of thick mud. He was over the valley.

~

At the helm of his lobster boat Jack Merrill yawned and scratched his beard, then draped his hand back over the steering wheel and looked at the cabin clock. It was a few minutes past 6 A.M. Jack seldom beat Bruce out in the morning, even though the twin-turbo diesel engine aboard Jack's boat, the *Bottom Dollar,* cranked out nearly two hundred more horsepower than the *Double Trouble*. This morning Jack would be even later than usual, because he had a task to accomplish before tending his traps. But given how worried he was about the lack of lobsters, it was a job he had to do.

A flash of reflected sunlight caught Jack's eye. He nudged the wheel to starboard, aiming his bow toward a white wedge on the horizon. Reaching overhead, Jack dialed his VHF marine radio to the hailing channel. He plucked the microphone from its clip and cleared his throat.

"This is the *Bottom Dollar*, calling the R/V *Connecticut,*" Jack said into the mike, his voice gravelly. From a loudspeaker by Jack's ear the response blasted back.

"This is the R/V *Connecticut,*" the voice said. "Go ahead."

Jack winced and turned the volume down.

"Good morning," he responded. "Is Bob up?"

"Yes. He's expecting you."

Fifteen minutes later Jack throttled back, twirled his wheel, and peered up at the ship that loomed above his boat. "R/V" stands for "research vessel," and the *Connecticut,* operated by the Marine Sciences and Technology Center of the University of Connecticut, was a state-of-the-art platform for the study of undersea life. Her bridge rose from behind her soaring bow like the control tower of a small airport, and her aft deck was equipped with a variety of machinery, including a gray A-frame crane for launching submersible equipment off the stern. Crew members wearing flotation vests and carrying walkie-talkies deployed rubber bumpers from the *Connecticut*'s rail. Jack maneuvered the *Bottom Dollar* to the side of the ship with forward and reverse thrusts of his propeller.

From inside the *Connecticut*'s superstructure a compact man strode on deck. His name was Robert Steneck, and he was a professor of marine science at the University of Maine. He was smiling.

"Hey, Jack!" Bob shouted.

Bob Steneck and Jack Merrill had been friends for fifteen years. Marine research and commercial fishing were two different worlds, and for nearly a century the relationship between scientists and lobstermen in Maine had been one of open hostility. But with many of New England's fisheries decimated by overfishing, Bob and Jack had joined forces in the hope of averting a similar disaster in Maine's lobster fishery.

"Good morning, Bob," Jack said. "I've got some numbers for you."

"Excellent," Bob said. He grinned and rubbed his palms together.

Bob pulled a notebook from his breast pocket. Jack produced a notebook of his own and read off several pairs of coordinates to the scientist—numbers he wouldn't have shared with his fellow lobstermen.

"That's where I've seen them," Jack said. "Big ones."

"Good," Bob said, jotting down the information. "We'll take a look."

The two men traded banter for a moment. Then Jack pulled away from the research ship, gunned his turbodiesel, and roared off toward his traps.

Bob stepped through a portal in the *Connecticut*'s bulkhead and strode through the ship's laboratory. Passing the smell of breakfast cooking in the galley, he mounted a steep stairway to the bridge. Surrounded by navigational electronics and hydraulic control levers, Bob studied a nautical chart and mapped out the coordinates Jack had given him.

"Two outcrops," Bob said, nodding. "Little underwater mountains." He sipped from a cup of coffee. "Just where you'd expect to find big lobsters."

Bob conferred with the *Connecticut*'s captain and put together a plan for the day. Bob was conducting a census of large lobsters. An average lobster in Maine waters required approximately seven years to grow to harvestable size. That was also about the age at which lobsters started to become sexually active, and lobsters old enough to copulate and reproduce were crucial to the health of the lobster population. If their numbers were dwindling, trouble could be in store for the lobster fishery. From the look of the catches this year, some feared trouble had already arrived. Bob wasn't so sure. With the help of lobstermen like Jack, Bob hoped the waters off Little Cranberry Island might provide some answers.

Younger lobsters tend to live in shallow water and can be studied using scuba gear. The older lobsters Bob was after on this trip were another matter. They had been known to live at depths exceeding two thousand feet, though most of them probably didn't venture much below several hundred feet. That was still too deep for comfortable diving with a scuba tank, so today Bob would remain aboard the *Connecticut* and send down the *Phantom* instead.

The *Phantom* was a submersible robot, referred to by the technicians who took care of it as a "remotely operated vehicle," or ROV. The *Phantom* belonged to the National Undersea Research Program of the U.S. National Oceanic and Atmospheric Administration. In the past, NURP's fleet of underwater robots had dived in exotic locales off Russia, in the Great Lakes of Africa, and at the North and South Poles. But NURP had granted Bob use of the *Phantom* for a mission closer to home: for the next ten days the robot would be stalking lobsters off the coast of Maine. Armed with searchlights, video cameras angled both forward and down, four whirring propellers, and a pair of lasers, the *Phantom* was likely to dominate an encounter with any lobster, no matter how large and antagonistic.

Or so Bob hoped. A few years back he'd been aboard a nuclear submarine owned by the U.S. Navy, cruising the sea floor off the continental shelf, when the sonar operator had reported a target at two hundred meters. Bob had slipped into the cramped observation module belowdecks. There, through a six-inch-thick glass portal, he'd been faced with the largest lobster he'd ever seen. She was a four-foot-long female, probably weighing thirty or forty pounds. She had turned toward the submarine and defiantly raised her claws.

~

The valley might be empty like everywhere else, but it was worth a try. Bruce Fernald had caught lobsters there in the past.

"O*kay*!" Bruce said to Jason, emphasizing the second syllable. "Let's get a pair on the rail." The boat rolled and Bruce grimaced. "It would have been easier to do this yesterday, when it was flat-ass calm."

Jason agreed. With his legs spread wide on the pitching deck he strode aft, wrestled down a bundle of buoys, and tossed them forward. Then he pulled a trap from the pile in the stern and hefted it onto the port gunwale.

The trap was a hollow rectangle made of plastic-coated wire mesh and divided into sections—a "kitchen" and one or two "parlors." A bag of bait went in the middle of the kitchen, strung between a pair of horizontal, outward-facing funnels knit from twine. Each funnel creates a ramp ending in a hole. Lobsters have an easy time walking up the ramp and through the hole; finding the hole and getting back out is more difficult.

Another funnel inside the kitchen led to the parlor, a compartment designed to hold the lobsters until the trap was hauled up. By law lobstermen are required to fit the parlor with rectangular vents through which little lobsters can escape. The vents are made of buoyant plastic and are attached to the wall of the trap with steel rings designed to corrode slowly in salt water. Should the trap's buoy rope get cut and the trap lost on the bottom, the rings will eventually disintegrate and the vents will float free, exposing a wider opening through which a lobster of any size can escape. Most traps are outfitted with several bricks, which help them sink quickly and stay in one place on the bottom. At a length of three or four feet and weighing forty pounds, a lobster trap is a hell of a thing to heft around. And if it snags a fisherman on its way overboard, it can drag him straight to the bottom.

Jason turned to retrieve a second trap while Bruce opened the first trap and extracted two coils of rope. In the Gulf of Maine, billions of tons of water flow in and out of the bays along the coast every day as the tide follows the tug of the moon. These hurrying seas are so strong that Bruce had to use rope twice as long as the water was deep, because anything shorter would be dragged under. The buoys were shaped like bullets, streamlined to offer less drag against the currents on the surface, and the ropes themselves were specially designed. The first coil, from the buoy to the halfway point, contained lead filament so it would sink, keeping it clear of the propellers of passing boats. But the second coil, from the halfway point to the bottom, was buoyant polypropylene. It would rise from the trap and float safely above abrasive rocks, even as the tide yanked it back and forth.

Jason hefted the second trap onto the rail next to the first. In water this deep, attaching only one trap to each buoy would be a waste of rope. Bruce had decided to set his thirty-three traps in fifteen pairs, plus one group of three traps at the end—a triple.

Jason opened the second trap and slid out another coil of the buoyant rope, which Bruce tied onto the main line near the first trap. Bruce then tied the main line to a buoy painted with his signature colors: white, black, and fluorescent red. Finally, he coiled through sixty feet of line and tied on another, unpainted buoy—called a toggle—which would spend most of its time underwater but would help keep the surface buoys accessible in the stiff currents.

Bruce rechecked the line, then glanced out the open panel in the windshield to ensure that the boat's bow was still pointed into the waves. If he let the *Double Trouble* get sideways to this sea, traps might start tumbling overboard when they weren't supposed to. Like most lobster boats, the *Double Trouble* was fitted in the stern with a mast and boom rigged with a triangle of canvas called the riding sail. Normally, the force of wind against the sail would temper the rolling of the boat and swing the stern downwind. But at the moment, the *Double Trouble*'s riding sail was furled and lashed to the mast to save deck space.

On the seafloor beneath the *Double Trouble,* the underwater valley was wide. If there were lobsters in it, Bruce guessed, they would be foraging along the edges. He would set eight of his pairs down one side of the valley, and another seven, plus the triple, back up the other. Each line of traps he referred to as a "string." But before he could drop the gear overboard, the boat would have to be properly positioned. The tide was ebb, flowing away from the coast at a brisk clip, so somewhat like a bombardier Bruce would have to drop each trap northeast of its target and let it sail southwest with the current as it sank.

"Hold on to those," he said to his sternman, spinning the wheel and gunning the engine. Bruce was staring at the GPS plotter when a wave shook the hull and the boat leaped into the air. Jason tightened his grip on the traps. A split second later the boat crashed down and a burst of spray splattered like machine-gun fire across the windshield. Half of it flew through the open windshield and slammed Bruce squarely in the face and chest.

"Whoa!" Bruce yelled, eyes wide. He growled and throttled down. Reaching for his waterproof jacket, Bruce caught Jason trying to suppress a smile, and both men laughed.

"There's just no need," Bruce said, invoking a phrase he might as well have patented, "of this unnecessary bullshit."

He pulled the window shut and switched on the Clearview, a circular plate of glass in the windshield that spun at eighty revolutions per second—fast enough to fling off oncoming walls of seawater instantly. He glanced at the GPS again, then gave Jason the signal to throw.

Jason turned the tail trap perpendicular to the gunwale and gave it a shove. As the trap splashed into the water he leaped nimbly backward, eyes riveted on the pile of rope at his feet, which was now playing out in a blur of flying coils.

When rope runs off a moving lobster boat it is reluctant to leave and will flail across the deck until it finds the point of exit that is farthest aft. Over the years, so many miles of rope had run off the *Double Trouble*'s decks that a deep groove was worn in the corner of her stern. But today her deck was piled with gear, and a rope flailing aft could cause mayhem. To coax the rope into the water sooner, Bruce had planted a piece of iron pipe upright in the gunwale, like a fence post. The rope was now flinging itself up from the deck, hitting the pipe, and falling overboard amidships.

Another wave hit the starboard bow and the *Double Trouble* rolled on her beam, the port gunwale sinking toward the water. Jason leaned back and held the head trap against his chest to keep it from sliding into the sea too soon. In the same instant the outgoing rope happened to flip over the top of the iron pipe.

The boat quickly righted herself, but now the rope was running overboard behind the pipe instead of in front of it. In seconds the coil on deck would be spent and the rope would yank the head trap aft inside the boat, slamming it into the stack of untethered traps in the stern and probably dragging some of them overboard. If Jason was lucky, the head trap would knock him out of the way as it passed. If he was unlucky, he could end up mashed between traps on his way into the water.

In four quick movements, Bruce used his right hand to flip the throttle to idle, throw the gear handle into reverse, and slam the throttle wide open again, while with his left hand he lunged for the bridle of the head trap to help Jason hold it aboard. The boat shook violently in protest and the water around her stern frothed. As the *Double Trouble* slowed to a halt, Bruce spun the wheel to port and with a burst of forward power swung the stern away from the submerged trap line that was trailing behind the boat. He had averted one crisis only to invite another—tangling the rope in his propeller.

Forty minutes later all thirty-three traps were in the water and the *Double Trouble*'s decks were clear. Jason pulled down his overalls and urinated onto the deck, then hosed it off, washing a mixture of pee, grime, and sun-dried periwinkles out the scuppers in the stern. Bruce plucked a fresh blueberry muffin from his lunch bag. The night before, Bruce had put on his best pouting face, and Barb had agreed to make the muffins. She knew from experience how miserable it could be out on the water.

While Jason struggled to open a Pop-Tart with his fish-oily hands, Bruce switched on the radio and tuned it to the oldies station. It was nearly 8:00 A.M. He set a course for the first of the three hundred traps he planned to haul that day. The traps had been sitting on the bottom for four days. Maybe there would be lobsters in them.

Bruce turned to Jason and grinned.

"I guess that could have been worse."

Jason nodded. "Yup."

~

The R/V *Connecticut* was hovering over the first dive position of the day. The crane pivoted off the stern, dangling the *Phantom* above the water by its tether. A technician hit a lever and the crane's winch creaked into action, lowering the robot into the sea. A voice from a loudspeaker crackled across the deck.

"ROV in the water."

Bob Steneck ducked into the *Phantom*'s command room. His eyes took a moment to adjust to the darkened scene within. A bank of video screens, computer keyboards, and racks of electronic equipment ran floor to ceiling through the narrow compartment. Sonar pings sounded, overlaid with radio communications between the command module and the bridge. In front of one screen sat the *Phantom*'s pilot. Next to him were a copilot, an engineer, and one of Bob's research assistants, their eyes glued to the screens. Off to the side, monitoring a video screen of his own, sat Carl Wilson, a sturdy young man with tousled blond hair. Carl was the chief lobster biologist at Maine's Department of Marine Resources.

"Hey guys," Bob chirped, perching next to his assistant, "what's our depth?"

"Just coming up on eight-zero," the pilot answered, steering the *Phantom* toward the bottom with a pair of joysticks. The copilot monitored the position of the robot relative to the ship. A breeze on the surface could nudge the *Connecticut* off the diving position and drag the robot backward by its tether. Following instructions from the robot's copilot, the *Connecticut*'s captain made constant corrections with pulses from the ship's bow and stern thrusters. On the video monitors, a rain of plankton gave way to a landscape of pebble fields and small boulders.

"Bottom in sight," the pilot radioed to the bridge. "Depth, one-zero-four."

Sea anemones grew like stalks of broccoli on the rocks. Small fish darted among a variety of bottom-dwelling sea life, including mussels, scallops, and starfish. Crabs lumbered across the sediment. Between rocks were nooks and crannies of the sort that Bob knew lobsters sought for shelter.

"This looks like a high-rent district," Bob said. "Let's start here."

Bob's research assistant switched on the video recorder and noted time and depth on a clipboard. The pilot set the *Phantom* onto a "transect"—a straight-line run of one hour in one direction, which generated data that was more statistically useful than random searching.

The *Phantom* glided over the gravel for several long minutes without encountering anything of interest. Then, in the distant gloom, Bob thought he saw the tip of a lobster's antenna protruding from behind a rock.

"There's one," Bob said, pointing. "Between those two boulders. Let's see if we can encourage him out of there."

The pilot pressed his joystick and the *Phantom* entered a slow-motion dive. The robot nudged the boulder and the lobster antenna twitched. Sure enough, when the pilot backed the robot away, the lobster emerged from its hiding place to investigate the intruder. It strutted forward, claws extended and antennae whipping the water.

If the lobster had been able to see the robot hovering overhead it might have been unnerved. The eyes of a lobster can detect motion under low-light conditions but don't discern much detail, especially when faced with floodlights. Lobsters are, however, equipped with sensitive touch receptors, in the form of their two long antennae and thousands of minute hairs protruding through the shells of their claws and legs. Like houseflies, lobsters can also taste with their feet. But a lobster's most acute sense is its ability to smell. A smaller pair of two-pronged antennae, known as antennules, contain hundreds of chemical receptors that give lobsters most of their hunting and socializing skills. But the *Phantom* didn't emit a recognizable scent. Uncertain, the lobster turned from side to side.

"That's it, baby," Bob cooed, leaning back in his chair. "Work the camera."

Bob wanted a side view in order to get a size measurement. If the *Phantom*'s pilot circled, the lobster was likely to pivot with the robot, claws at the ready. Instead, the pilot feigned retreat by backing up. Concluding that the threat had passed, the lobster turned to walk away, exposing its flank.

"Paint him with the lasers!" Bob exclaimed, scooting to the edge of his seat.

A pair of parallel laser beams hit the lobster squarely on its shell, providing a gauge of the animal's length. Satisfied, Bob sat back. The pilot recommenced the transect. Shortly Carl Wilson squinted and pointed to a corner of the screen.

"Is that another one over there?"

"Yeah, and he's running away," Bob said. "Hit the afterburners!"

The pilot changed course, and the *Phantom* slowly gained on the lumbering lobster. It was a hulking animal, barnacles growing on its shell. The big lobster turned, faced the *Phantom* head-on, then lifted its claws wide and ran directly at the robot.

"You're going to lose," said the pilot.

At the last second the lobster seemed to reach the same conclusion and backed off.

~

Bruce Fernald finished hauling his traps early. It had been another miserable day. Bruce and Jason had emptied and rebaited nearly three hundred traps for a measly seventy-five lobsters. By tradition a sternman's earnings were a fixed share of the catch. Today Jason had made the mistake of calculating his hourly wage. Bruce had made the mistake of pondering the pair of college-tuition payments he was making for his twin sons. He'd done his part to repopulate the world. Why weren't the lobsters doing theirs?

Bruce and Jason were scrubbing the boat down on their run back toward shore when the marine radio crackled.

"This is the R/V *Connecticut* calling the *Bottom Dollar*. You on there, Jack?"

Bruce recognized Bob Steneck's voice and turned up the volume to listen.

"Yeah, this is the *Bottom Dollar*. Go ahead."

"Hey, Jack. It's Bob. How's it going?"

There was a moment of silence before Jack answered.

"Ah, it's not looking so good out here. Did you get a chance to check out those spots I gave you?"

Bob explained that partway through the day a computer had malfunctioned in the *Phantom*'s command module, delaying the dive schedule.

"Unfortunately, I won't have time today," Bob said. "But I'm going to try to hit them next week, on our way back from Canada."

"That's too bad," Jack said.

"Yeah," Bob said. "Anyway, good luck with the rest of your day."

The radio went quiet. Bruce shook the soap from his brush and scanned the water for the *Connecticut*. He could make out the white wedge of her bow steaming in from the west. He altered his course twenty degrees so the *Double Trouble*'s path would intersect the *Connecticut*'s.

A few minutes later Bruce throttled down as his lobster boat pulled up to the research ship. Bob Steneck and Carl Wilson talked with Bruce across the trough of seawater splashing between the two craft.

"Did you clean up today, Bruce?" Carl shouted, smiling.

Bruce groaned.

"Hardly caught a thing," he said. "Thought I'd stop by and complain."

The men laughed. Then Bruce grew serious.

"So far this is the worst season I can remember."

Bob nodded. "I've been talking to fishermen all along the coast," he said, "and it's the same story everywhere. No one's catching any lobsters."

"It's downright grim," Bruce said. "How's it look on the bottom?"

"We did see some lobsters today," Bob answered.

"I sure as hell would like to know what's going on down there," Bruce said, shaking his head. "When you figure it all out," he added, only half joking, "let me know."

Bruce backed his boat away from the research ship, leaving a frothy wake. He threw the scientists a salute, then punched the throttle and set a course for home.

A few minutes later the radio aboard the *Double Trouble* crackled once more.

"*Bottom Dollar,* you still on there, Jack?" It was Bob again.

"Go ahead," came Jack's voice over the speaker.

"I don't know if it makes any difference to you where you're fishing, but I just told Bruce that over here we saw some lobsters on the bottom."

"Is that right," Jack responded. "Throw a few in my traps, will you?"

"Yeah, right." Bob laughed.

The voice of another local lobsterman interrupted the conversation. "You saw lobsters?" he said. "Where the hell are you? Stay right there, I'm on my way."

PART ONE

Trapping

1

A Haul of Heritage

The oceans of the earth abound with lobsters. Lobsters with claws like hair combs sift mud in offshore trenches. Clawless lobsters with antennae like spikes migrate in clans in the Caribbean and the South Pacific. Flattened lobsters with heads like shovels scurry and burrow in the Mediterranean and the Galapagos. The eccentric diversity of the world's lobsters has earned them some of the most whimsical names in the animal kingdom. There is a hunchback locust lobster and a regal slipper lobster. There are marbled mitten lobsters, velvet fan lobsters, and even a musical furry lobster. The unicorn and buffalo blunt-horn lobsters inspire admiration; the African spear lobster, the Arabian whip lobster, and the rough Spanish lobster demand respect.

Nowhere in the world, however, is the seafloor as densely populated with lobsters as in the Gulf of Maine. Though a less sophisticated creature than some of its clawless counterparts, the American lobster, scientific name *Homarus americanus,* is astonishingly abundant.

But at five o'clock on a September morning in 1973, the young Bruce Fernald didn't know that, and he wasn't interested.

"Hey, Bruce." The door opened. "Come on, son, get up. We're going fishing."

Bruce groaned, rolled over, and cracked open an eye. Still dark. Jesus. Almost four years in the navy, riding nights away in the bunk of a destroyer, rounding the Cape of Good Hope in forty-foot seas, and what happens the first time he tries to sleep in his own bed back home? His father wakes him up before dawn to get in a boat.

Sure, Bruce thought as he yanked on his socks, when I was fourteen I hauled traps by hand from a skiff, like every other kid on Little Cranberry Island. Does that automatically make me a lobsterman? The world was big and in the navy Bruce had sailed all the way around it. He wasn't certain he wanted to condemn himself to the hard life his forefathers had endured, hauling up what the old-timers called "poverty crates" full of "bugs."

But Bruce's first day of lobstering with his father turned out to be lucrative enough to warrant a second day, and after that a third. As autumn settled over the island the days aboard his father's boat became weeks. At the helm was Warren, his dad, and on the stern was the name of his other parent—*Mother Ann*. Bruce stuffed bait bags with chopped herring. He plugged the lobsters' thumbs with wooden pegs to immobilize their claws so they wouldn't rip each other apart in the barrel. He coiled rope. He hefted the heavy wooden traps. And he observed his father at work.

Some of Warren's white-and-yellow buoys followed the shoreline like a string of popcorn. Warren knew just how close he could get to the rocks without endangering the boat, and he showed Bruce how to line up landmarks and steer clear.

Some of Warren's buoys bobbed in ninety feet of water, running in a line east to west half a mile from the island. Unwritten rules along most of the Maine coast governed just how far a fisherman could go before he was setting traps in someone else's territory. Bruce watched where his father went and memorized the landmarks that would keep him close to home.

Come November, Warren and Bruce were hauling traps in water twenty fathoms deep—120 feet—a mile south of the island in open sea. It was cold, especially when the breeze picked up and blew spray in Bruce's face.

"Okay, son, where are we now?" Warren asked, bent over a tangle in the rope.

Bruce, his hands numb, glanced up to see which of the mountains of Mount Desert Island loomed over the lighthouse on Baker Island, half a mile southeast of Little Cranberry. Depending on how far to the east or west the *Mother Ann* was positioned, the lighthouse would line up with a different hill.

"Cadillac," Bruce answered.

Cadillac Mountain, like the automobile of the same name, honored the first European settler in these parts. In 1688 a small-town French lawyer swindled a land grant to Mount Desert Island from the Canadian governor. He invented the aristocratic title "sieur de Cadillac" for himself and lorded it over the uninhabited island with his new bride for a summer. Bored, he soon retreated inland to found a trading post called Detroit. The Cadillac car still bears his fake coat of arms on its hood. The lobstermen of Little Cranberry had put Cadillac's legacy to their own use. Like the other hills of Mount Desert, his mountain rising from the sea was a map to the treasures under the waves.

In a more literal sense too, Warren and Bruce were fishing on Cadillac Mountain—or at least on pieces of it—and that was what made these waters hospitable for lobsters. Starting a few million years ago, sheets of ice had rolled down from the Arctic for eighty thousand years at a stretch, interrupted by brief warm spells of ten thousand or twenty thousand years. During the most recent ice age the glaciers had scraped up stone from all over Maine and carried it south, carving away at the pink granite of Mount Desert Island on the way. The glaciers had pressed on for another three hundred miles before grinding to a halt, encrusting the Gulf of Maine and the continental shelf in ice as far south as Long Island.

When the glaciers melted fourteen thousand years ago they unveiled the sensuously sculpted hills and valleys that now constitute Acadia National Park. The glaciers also left behind vast fields of debris—boulders, cobble, pebbles, and gravel. Glacial runoff sorted the finer sediments into beds of sand or muddy silt between ledges of hard rock. Sea levels rose, filling in the convoluted coastline and creating islands, bays, inlets, and in the middle of Mount Desert, the only true fjord on the east coast of the North American continent. Underwater, this terrain of rocks and sediment became the perfect habitat for lobsters. It was an intricate rangeland that Bruce would have to learn by charts, depth sounders, compass points, and intuition rather than by sight. The more he thought about it, the more this seemed a task that might warrant a lifetime.

~

Bruce's great-great-great-grandfather Henry Fernald had settled on the island next door, Great Cranberry. But with a paucity of women there, his three sons had rowed the half mile across the water to Little Cranberry in search of mates. They wooed local girls, married, and settled on the smaller island. When they'd had enough of home life they jumped in their dories and went to sea.

From their boats the Fernalds had lowered lines weighted with a chunk of lead. A clank when it hit bottom meant rock, a thud meant sand, and nothing meant mud. They marked off the depth in fathoms and rowed around feeling where the rock went, then gave each underwater feature a name honoring its shape, characteristics, or the man who found it: Bull Ground, Moose Ground, Mussel Ridge, Tide Hole, Smith's Shoal, Poag's Piece, or George Hen's Reef—the last named by Bruce's great-great-great-uncle George Henry. And in the spring they returned to set their traps.

Lobster traps were a newfangled technology when Bruce's ancestors started using them. Lobsters had been caught by various methods for a long time before that. European explorers dragged up Maine's greenish brown lobsters from shallow water with hooks. The animals looked familiar because European waters were home to the American lobster's nearly identical twin, the bluish black *Homarus gammarus*. Although the two species have evolved separate colors of camouflage, both turn red when boiled in the pot. During cooking, protein molecules in the shell bend into shapes that absorb different wavelengths of light and end up reflecting red.

These two species also share something akin to a secret undergarment of the brightest blue. If extricated, proteins from the shell of the mostly black *Homarus gammarus* can be grown into brilliant blue crystals, and every so often a specimen of the mostly brown *Homarus americanus* undergoes a rare genetic mutation that unveils its stunning inner indigo. American lobsters that don't get enough calcium in their diets can fade from brown to blue too, but of a less vibrant hue. Genetic mutations of yellow, white, calico, and even red also turn up in living lobsters, and very occasionally one is caught that is half-and-half—the line down the middle of its back as straight as a ruler.

It was from the European *Homarus gammarus* that the name "lobster" originated. The Old English version of the word, "loppestre," is probably related to loppe, meaning spider. But the original derivation likely goes back to the Latin *locusta*. Pliny the Elder, writing in his *Natural History* during the first century AD, observed that when a lobster was surprised, it seemed to "disappear with a single leap or bound as a locust or grasshopper might do," and so he used the term *locustæ*—locusts of the sea. With the lobster's obvious resemblance to an insect, the name stuck. Until the English word was standardized, writers used spellings as various as "lapstar" and "lopystre" to refer to the crustacean.

Historians of New England often note that early settlers considered lobster a kind of junk food that was fit only for swine, servants, and prisoners. These claims may be exaggerated. But storms could blow lobsters onto beaches by the hundreds, making them a convenient source of feed or fertilizer for coastal farms, and most scholars agree that lobster was generally considered a low-class dish for human consumption. After their first winter in Plymouth, a group of Pilgrims on an expedition to what is now Boston Harbor gladly helped themselves to fresh lobsters that had been piled on the beach by Native Americans. By the following year, however, the leader of the

Pilgrims, William Bradford, reported shame at having to serve lobster in lieu of more respectable fare.

By the seventeenth century, the word "lobster" had even developed a derogatory usage in speech—calling someone a lobster was like calling him a rascal. One English source from 1609 gives an example: "you whorson Lobster." During the American revolution, the word was a put-down for British redcoats, and in American slang of the late 1800s it was used to call someone a dupe or a fool.

Despite these connotations, fishermen along the New England coast ate lobster, though primarily out of economic necessity—the fish they caught were too valuable at the market to consume, while lobster was nearly worthless. Gradually the lobster's status improved, and its meat became desirable fare for well-off urbanites. By the early nineteenth century, American fishermen were catching lobsters commercially with a type of net hanging from an iron hoop and shaped like a cauldron—one origin of the term "pot," still used today to refer to a trap. Traps of wood and twine were far more efficient than nets and caught on in New England in the 1840s.

For Bruce Fernald's forefathers, building twenty or thirty traps could take all winter. The men hauled spruce from the forest and sawed it into sills, then stripped green branches and soaked them in a round washtub to make the arched bows that gave the traps their curved tops. The women who had been foolish enough to marry these men sat by the stove knitting mesh funnels and bait bags from twine. Then the men boiled vats of coal tar and cooked the twine to fortify it against decomposition. While the tar was hot they measured lengths of rope made from Manila hemp or sisal plants and cooked them too.

The buoys they carved from tree trunks, each man painting his floats a signature color. Just before they set a trap they loaded it with beach stones so it would sink; after a week or two the wood would be waterlogged enough that they could remove some of the rocks. And then on a good day each man piled as many traps as he could into a dory, rowed out to where he thought the lobsters were, and threw the traps overboard. When he hauled a trap back up and found lobsters in it, he noted its location and reminded himself to lie about it when he returned to the island.

For a hundred years the Fernalds mostly set their traps on rocky bottom, where they believed the lobsters liked to hide. Occasionally a storm would churn up the sea and drag the wooden traps to new locations, often off the rocks and into mud valleys several miles away. After the storm the Fernalds would head out in their boats to search for their gear. When they found a trap they were relieved, whether it contained lobsters or not. Still, as the years passed, the lobstermen began to notice an odd phenomenon. Sometimes the traps they retrieved from the mud seemed to contain more lobsters than the traps that had stayed on the rocks. Perhaps the animals had been frightened into the muddy valleys by the raging currents of the storm.

By the time Bruce Fernald was fishing with his father, a new theory had developed. Perhaps the lobsters used the rocks for hiding, but the mud for migrating. It was a theory Bruce grew increasingly eager to put into practice for himself.

Jack Merrill's family lived in suburban Massachusetts. His parents brought him to Little Cranberry Island before he was a year old, and Jack spent his boyhood summers entranced by the island's rocky beaches, its stands of spruce, and the scent of salt in the air. One day he eagerly accepted an invitation from an old-timer to go lobstering. At 6:00 A.M. the young Jack nearly lost his breakfast walking past the bins of rancid bait on the wharf. But staring straight ahead, he held his breath and made it aboard the old wooden boat and out onto the sparkling sea.

"Nature has a way of separating the men from the boys," the old-timer said, pouring himself coffee from a thermos and soaking up the sunrise.

For Jack, the end of each summer on Little Cranberry, and

the subsequent reversion to suburban life, was torturous. He vowed to make something of his affection for Maine's craggy coast and wide-open ocean. His ancestors on his father's side had come from Maine, and his great-grandfather had been a governor of the state. As he neared adulthood, Jack grew certain that he wanted to return to these family roots.

This dream nearly became the death of him. Jack taught marine ecology for several summers at an outdoor adventure camp in Maine. Once, he and a group of campers were sailing a pair of thirty-foot open boats out of Hurricane Island when an October gale whipped up enormous waves. Both boats were swept out to sea. Jack and his fellow sailors rode the storm through the night, bailing to stay afloat. The gale subsided and in the morning the Coast Guard found them, chilled and exhausted. Jack figured that if he could survive that, he could survive commercial fishing.

By the age of twenty-one, Jack was back on Little Cranberry Island and ready for a job as sternman with Warren Fernald. During his first week aboard the *Mother Ann,* Jack stood by with a can of claw plugs when each trap broke the surface. Often the traps were loaded with lobsters flapping their tails. Warren would open the trap and reach among the snapping pincers. He tossed most of the animals overboard without a second glance. Several he kept only long enough to slap a brass ruler on their backs before throwing them back into the water too.

The trap would be empty and Jack would have yet to change the bait. He would hurry to unwind the spent bag and hang a fresh bundle of herring in the trap. Warren would tie the door shut and shove the trap overboard. After an hour of this routine Jack would steal a glance in the barrel of keepers. More often than not, he could still see the bottom of the barrel. Warren threw more lobsters overboard with each trap he hauled.

"This is nuts," Jack said under his breath. Another trap came over the rail, full of shiny lobsters that would go back into the sea.

"Too bad we can't just keep all these," Jack muttered.

Warren thought for a minute.

"You want to be a lobsterman?" Warren asked.

"I don't know. Yeah, maybe."

"You want to keep on lobstering after you start?"

"Yeah, sure."

"Well, you can't catch everything and expect it to continue," Warren said, turning the brass ruler over in his hands. The ruler, which lobstermen called the "gauge," enforced a minimum-size law that had been in effect in Maine since 1895. "Throw back more than you catch and, why, there's always going to be something there tomorrow."

As if to punctuate his point, later in the string of traps Warren turned a female lobster on her back and showed her to Jack. Glued to the underside of her tail were thousands of pine green eggs. Warren reached for his fish knife and cut a quarter-inch triangle out of the lobster's tail flipper, then slid her back into the sea. He had just marked her as a breeder by bestowing her with a "V-notch," so-called because the triangular cut was shaped like a V. If caught again, the lobster would be illegal to sell whether she was carrying eggs or not.

Jack was getting used to the idea of returning "shorts," "eggers," and "V-notchers" to the sea when a trap came over the rail containing a mammoth lobster that could have crushed a man's wrist in its claws.

"Now, that's a handsome fellow," Warren said, noting the more muscular claws and narrower tail that indicated a male. Females have wider tails to accommodate their eggs.

With a couple of skillful tugs Warren extracted the creature and held it up for admiration. Then he dropped the lobster overboard with a splash. Jack leaned over the rail and glimpsed the animal pulsing its tail and retreating into the depths.

"Too big," Warren said. In addition to a minimum size, his brass ruler delineated a maximum size, another law that Maine had pioneered, in 1933.

"The oversize lobsters are our brood stock," Warren explained. "We protect the big males so they can mate with the big V-notched females that produce the eggs."

Warren hauled up another trap and found a female without eggs. But close examination revealed the remnant of a nick in her tail.

"A notcher that's shed her old shell," Warren said, showing the lobster to Jack. "You almost can't make out the notch anymore." Warren cut her a new notch before tossing her overboard.

He wasn't just trying to repopulate the waters around Little Cranberry with lobsters. By now Warren and Ann had six children. Their sons were joining the ranks of the island's lobstermen, and Warren had taught them how to protect lobster eggs just as he was teaching Jack. If Jack's interest took root, there might be another lobsterman living on the island, one more steward for Little Cranberry's female lobsters.

Jack stuffed another bag with bait. He looked up from the tub of putrid herring and gazed at the sparkling waves lapping at the edge of the island. Warren's lessons on lobster fertility had sparked in him a new appreciation for the possibilities of procreation.

"There's a future here," Jack whispered.

2

Honey Holes

Pa's Pride was a creaky little boat. Bruce Fernald nudged her throttle to speed up the hydraulic trap-hauler and prayed she'd hold together. The boat had been Warren's when Bruce was a boy. A year had passed since Bruce had returned to Little Cranberry Island from the navy, and he supposed it was appropriate that now, as the eldest son, he was the boat's captain. But mostly the name *Pa's Pride* reminded Bruce that he was twenty-three years old and six thousand dollars in debt. No one was going to be proud of him if he didn't start catching some lobsters.

Just trying to keep *Pa's Pride* in one piece was hard enough. A storm had bombarded the harbor with screaming winds, and *Pa's Pride* had broken loose from her mooring. She'd banged up against the wharf, and Bruce had smashed open a dock window in the attempt to jump aboard and save her. For fear of being pulverized between the boat's hull and the wharf pilings, he'd given up and waited for her to slip ashore instead. With help from Jack Merrill and several other lobstermen, Bruce had dragged *Pa's Pride* up the beach with an old backhoe.

Saving the boat would have been pointless, though, if Bruce's traps came up empty. Bruce had dropped these traps overboard a week ago. He'd watched the stylus on the old-fashioned Fathometer burn a squiggly line onto a rolling sheet of paper like a lie detector. If the Fathometer told the truth, then the traps should have sunk into a muddy canyon. Bruce had gambled that lobsters would be migrating through it on their way offshore.

Outwitting the lobsters was only part of the battle. A good-natured competition simmered between Bruce and two of his younger brothers. Mark Fernald and Dan Fernald were the kind of island boys who looked naked without a lobster trap in their hands. When Mark was but a baby, his first encounter with lobsters had been life-threatening. When the lobsters Warren had brought home for dinner started to boil, steam wafted around the kitchen. Mark screamed and stopped breathing, his throat constricted by swelling. He suffered from a rare allergic reaction to proteins in lobster muscle and would never taste a lobster in his life.

That didn't stop Mark from setting out to catch more lobsters than any Fernald in the history of Little Cranberry. By the age of twelve he was hauling five traps from a rowboat. He found a sixth trap washed up on the beach, which he dragged home, repaired, and carted down to the harbor in a wheelbarrow. He hung a bait bag full of herring inside, rowed a quarter mile from shore, tied a buoy onto the buoy line, and slid the trap overboard. As he watched it sink he wondered why the buoy and line had stayed aboard the boat, then realized he'd forgotten to tie the line to the trap. Mark went on to amass hundreds of traps and it was a mistake he would rarely repeat. He was always curious to see what his traps would haul up, and he frequently shifted his gear around to test different types of terrain. One year Mark built a trap the size of a small car just to see what it would catch. It was so big that to complete its construction Mark had to climb inside the trap himself.

Dan was eager to pit his skills against the sea too, and by the age of seventeen was lobstering from an eighteen-foot boat. Bruce was still in the navy when Dan bought *Pa's Pride* from Warren, in April of Dan's senior year in high school. The evening he graduated, Dan went straight to bed so he could rise at 4:00 A.M. to haul his traps. Aboard *Pa's Pride* Dan developed a lightning-fast hand. With only a few flicks of the wrist Dan could empty a trap, rebait it, and set it back over the side before the boat drifted off target. Hitting the sweet spots on the bottom was an art, and all the Fernald boys dreamed of matching the uncanny talent of a man named Lee Ham.

When Lee Ham went lobstering it was almost as if he were making love to Mother Nature. Lee had a knack for planting his traps in the depressions in the seafloor, where lobsters liked to hide and hunt. He called these spots his honey holes. He caught the most lobsters of any of the Little Cranberry fishermen, and he made a profit on everyone else's lobsters because he owned the dock where the fishermen sold their catch. Warren and his Fernald cousins sometimes traded their knowledge of the bottom, but Lee kept his honey holes to himself. It was said that he knew every pebble in the shallows around the island and every boulder offshore. Lee would steer his boat in among the rocks by the beach, propeller churning just above the stones, and set his gear so close to the lobsters' hiding places that they had little choice but to enter his traps. He would steam far from land and drop his traps offshore with eerie precision, making rows of pit stops on the lobster highways of the deep.

What the Fernald boys couldn't match in their knowledge of honey holes, Mark made up for with sheer ambition and Dan with speed. Both prospered, and soon Dan needed a bigger boat. He'd heard about a new kind of material called fiberglass. In 1974, at the age of nineteen, Dan ordered a hull fashioned from the stuff and christened her *Wind Song*. She cost as much as a house.

"What if something aboard her needs fixing?" one of the old-timers asked Dan. "Where the hell are you going to drive a nail in her if she's made of plastic? You paid forty thousand dollars for a goddamn Clorox bottle."

Dan smiled because his boat would never need caulking, sanding, or painting. He kept hauling like lightning and sold *Pa's Pride* to Bruce. Now Bruce needed a few honey holes of his own.

~

Bruce had figured out the basics. Most adult lobsters around Little Cranberry Island spent the winter hunkered down twenty miles from shore, on the mud plains two hundred to three hundred feet underwater. It was so cold that they didn't move around much, but they were warmer there than near shore, where winter's bitter winds cooled the shallows quickly.

Come spring, the sun warmed the surface of the sea and the lobsters set out toward land. At the edge of the mudflats they probed for the fingers of rocky ledge that rose toward the Cranberry Isles like mountain ridges rising from the desert. Between the ledges were silt-bottomed canyons that wound toward the islands. But with the surface waters warming, the lobsters mostly avoided the canyons and sought hard rock and altitude. When they hit the rocky slopes they clambered up toward ridges that would take them higher still, and closer to the realm of men.

About five miles from Little Cranberry the lobsters cleared a lip of ledge and emerged onto a boulder-strewn plateau. After their steep ascent, they walked at a more relaxed pace, doing most of their moving at night and sheltering themselves in crevices by day. They'd risen to a depth of eighty or ninety feet, where the terrain, though flat, was littered with rocks that tested their ability to detour without straying off course. The lobsters paid little attention to the inhabitants they passed—sea urchins, starfish, snails, and sea cucumbers. Occasionally the lobsters hunted down one of these creatures for a meal on the road, but they were intent on covering ground. They seldom paused, even for the bait in a lobsterman's trap. The warmth of the islands beckoned.

The plateau ended abruptly. The lobsters crossed a trough of gravel and drew up against another steep rise. They gripped the craggy slope with their four pairs of legs and scrambled upward again, the water growing warmer as they drew closer to the sun. They were in the home stretch now—sixty feet deep, then fifty, then forty. The incline flattened and the lobsters began to sway with the swells rolling toward the beach. For the first time in months they basked in the heat of the shallows.

Having mounted the rocky table that formed the

Cranberry Isles, many of the lobsters made their way into a mile-wide cove surrounded on three sides by islands. If the lobsters had continued straight ahead, they would have emerged onto the gray cobble beach of Little Cranberry's seaward shore and climbed into the grass and up into the woods. Bruce, in his sleep, sometimes saw them there, hoards marching up the hill. He dreamed of setting his traps in the island's streambeds and in the drainage ditches along the road so he could catch them before they reached town.

In reality the lobsters stopped and fanned out to find hiding places in the glacial debris underwater. Scurrying among the rocks, the lobsters sought the warmest nooks they could find, sometimes just fifteen or twenty feet below the crashing surf. The lobsters had good reason to secure hiding places. By now it was early summer, and in the warm water their shells had begun to loosen.

A lobster's shell gives the animal all of its rigidity. Under the shell, the lobster is little more than jelly-soft flesh and floppy organs. The problem with this arrangement is that the lobster is constantly growing, while its shell is fixed in size. To get bigger a lobster must literally burst its seams, escape its old shell, and expose its vulnerable inner self to the hungry world while it constructs a new shell large enough to allow its body to expand.

The shell is composed of three layers. The outermost is a thin covering of proteins, lipids, and calcium salts. Underneath is a thicker matrix of proteins and the horny substance known as chitin, the same material that forms the exoskeletons of insects, as well as of arachnids like spiders and scorpions. The third layer is thicker still, and consists of a rigid, calcified outer portion that becomes softer toward the inner surface, like a suit of armor lined with padding. Underneath is a protective membrane, and finally the lobster's delicate skin.

If a lobster's bodily functions went unregulated, the animal would be in a constant state of shedding its shell and growing a new one. Such exhibitionism would make normal life impossible, so lobsters have glands inside the stalks of their eyes that release a hormone that inhibits molting. A combination of cues, including warming temperatures and longer days, constrains the production of the hormone and releases steroids that begin the molt cycle. Proof of the hormone's importance can easily be obtained, though the experiment is somewhat sadistic: cutting off the lobster's eyes induces the animal to shed its shell almost immediately.

In the weeks prior to molting, the lobster's skin cells enlarge and secrete the beginnings of an entirely new shell underneath the old one. Meanwhile, calcium drains out of the old shell and accumulates in a pair of bulbous reservoirs on either side of the stomach called gastroliths, to be recycled later.

When the lobster is ready to shed, it pumps in seawater and distributes it through its body, causing hydrostatic pressure to force the old shell away from the new one. The lobster remains mobile and active until the last minute, when the membrane that lines its old shell bursts and the animal falls over on its side, helpless and immobilized. After twenty minutes or so, the lobster's back detaches and the animal pulls its antennae, mouthparts, legs, and claws out of their former coverings, aided by a lubricating fluid. The most difficult moment comes when the lobster tugs its claw muscles out through the slender upper segments that form its wrists. Before molting the animal must diet away half the mass in its claws or risk getting stuck in its old clothes. Worse, because a lobster is an invertebrate, every anatomic feature that is rigid is part of the exoskeleton, including the teeth inside the stomach that grind food. The lobster must rip out the lining of its throat, stomach, and anus before it is free of the old shell. Some die trying.

When the lining of the stomach comes out, the gastroliths, containing the calcium reserves, are released. The lobster immediately digests the gastroliths to recycle some of the rigidity of its old shell. Centuries ago, gastroliths from lobsters and crayfish were commonly ingested by humans as medicine. In the 1700s, apothecary shops throughout the Russian Empire sold the little white balls to dissolve kidney stones, heal eye inflammation, and cure epilepsy. Gastroliths were collected from crayfish by catching thousands of the animals in the summertime, dumping them into pits, crushing them, and letting them decompose over the winter—the stench was said to be horrendous. In the spring the mess was washed down and the gastroliths sorted out with sieves. Demand for the medicine was so strong that fakes, formed from chalk, were common. For a molting lobster the gastroliths have a specific function. Their calcium bypasses most of the new shell and goes directly to harden the tips of the legs and the cutting edges of the mouthparts—the appendages critical for feeding.

Flexing the muscles of its abdomen, the lobster shakes off the old shell around its tail and is free. Again the lobster pumps itself full of water and expands, rapidly outstripping its former size. Soon it is able to stand, and its first priority is to use its newly rigid mouthparts to devour the husk of its former self, a convenient and nutritious source of additional calcium. What the lobster doesn't finish of its old shell it buries, perhaps to hide evidence of its weakness.

The minerals and nutrients the lobster absorbs are secreted throughout the new shell, which over the next few days thickens enough to allow the flesh underneath to shrink back to its actual size, making space for future growth.

The lobster has gained 15 percent in body length and 50 percent in volume. In the first five years of life lobsters undergo this hazardous routine about twenty-five times. In adolescence the rate decreases to about twice a year. Lobsters that have reached adulthood molt once a year on average, and increasingly seldom as they grow larger. Exactly how often, though, is difficult to calculate, because determining the age of a wild lobster has so far proved impossible. A fish contains a chronological record of its life in its otolith, a bone inside the brain cavity that grows in concentric layers like the rings of a tree. But lobsters are invertebrates. They toss their exoskeletons off with such maddening thoroughness that all trace of their age is erased.

Bruce once saw a lobster molt aboard his boat. The lobster

had mistaken a trap for a sanctuary and let its shell loosen, only to be hauled up after it was too late. Bruce slid the lobster into a bucket of seawater and watched its shell open. With heroic patience and some spasmodic flailing the lobster extracted itself and emerged wrinkled and soft as Jell-O. But by the end of the day it had restored its shape and grown larger. Coating its skin was a crinkly shell the consistency of cling wrap.

~

With the arrival of molting season Bruce learned to plaster the shoreline with his gear, steering his boat as close to the rocks as he dared and tucking his traps between boulders near the surf. A week or two after shedding, the lobsters' big new shells were strong enough to allow them to emerge. As hunger overcame them, they were lured from their hollows by the lobstermen's bait, and by midsummer the waters around the island teemed with new "shedders" that met the minimum legal size for capture in the state of Maine. They entered the traps in droves, and their meat tasted especially sweet.

When the lobsters came out of hiding Bruce hauled like a madman. But he also learned that he couldn't afford to keep all his gear in the shallows. Within days the shedders were combining their hunt for food with another imperative—the return to deeper water. Soon other lobsters, hiding farther from the islands, would shed and emerge too. The wave of molting would progress offshore through the autumn, often lasting until December, and the trick for a lobsterman wasn't so much to follow the lobsters as to anticipate where among the rocky boulevards and muddy valleys the animals were going to be. If the traps weren't already in place when the lobsters passed through, a fisherman had missed his chance.

Bruce knew that to pay off *Pa's Pride* he would have to test new locations on the bottom. Eyes glued to the Fathometer, he had run the boat due east, and after a mile the stylus had dropped. He'd circled, then run south and circled again. If he'd read the burn line on the Fathometer's roll of paper right, he'd found a muddy canyon 150 feet deep, heading straight out to sea. Bruce had set a string of traps just beyond the drop-off.

Now, a week later, he was hauling those traps back up. The first one broke the surface and Bruce shut off the hauler. He grasped the rope bridle to tug the trap onto the rail, but it was too heavy. Grabbing it with both hands, he seesawed it out of the water and saw that the trap was oozing a pungent brown goo.

"Congratulations," Bruce muttered through gritted teeth, "you've hauled up a load of stinking mud."

He wrestled the trap aboard and wiped the mud off in gobs, smearing his rubber overalls and the deck with slime. He leaned over and peered between the wooden slats of the trap. Inside was a glistening pile of shedders.

PART TWO

Mating

3

Scent of a Woman

At the age of twenty-five Jelle Atema had left the Netherlands on a plane bound for America. He had athletic good looks and a promising international career, and was a gifted flute player—a student of the French master Jean-Pierre Rampal. In high school Jelle's goal had been to sculpt himself into a Renaissance man. If he had known that in the United States his obsession would become the sex life of the American lobster, would he ever have left Europe?

Four years later it was best, Jelle supposed, not to dwell on such questions. He chuckled, dropped a chunk of fish into each lobster's tank, and surveyed his charges—half of them male and the other half female. Within moments the lobsters flicked their antennae at the scent of prey and began waving their feeding mandibles. In a few seconds his lobsters were gnawing on their food. It was astonishing to Jelle how fast the American lobster worked.

Each lobster lived in a thirty-gallon Plexiglas tank with flowing seawater. Jelle had spent the past several days setting up cameras and lights next to the tanks and fine-tuning his behavioral coding system for male-female interaction. All that remained was for one of the females to get undressed.

The basic reproductive cycle of the American lobster begins when the female sheds her old shell. This arrangement gives the female time between molts to mate, lay her eggs, and carry the eggs until they hatch before she sheds her shell again—a schedule made necessary by the fact that the eggs are attached to the shell and would be lost during a shed. Jelle had heard reports of American lobsters copulating even when the female's shell was hard, but most female lobsters seemed to mate when their bodies were soft, immediately after molting. Jelle's first female to get lucky would be the one that molted first. It was June and they were all on the verge of shedding.

As a graduate student in the Netherlands Jelle had studied how a species of small shark detected vibrations and electric fields in the water. While finishing his Ph.D. in 1966 Jelle had been invited to join a group of scientists at the University of Michigan who were studying how fish sensed chemicals. In Ann Arbor Jelle became entranced with the catfish, and how it used smell and taste. Underwater, smell and taste are, in a sense, simply two aspects of the same thing—the detection of molecules in water. The difference has to do with the organs employed and the purposes to which detection is put. Jelle's catfish used their sense of smell to socialize and select mates, identifying one another as individuals by means of unique odor signatures. They used taste to search for food, licking the water, as it were, to find dead flesh.

Jelle's research on the subject was getting under way when the prestigious Woods Hole Oceanographic Institution in Massachusetts offered to build him a seawater laboratory on Cape Cod. It was an offer he couldn't refuse, especially when Jelle learned that the institution wanted him to head up an ambitious new program: the study of chemical communication in the sea. There was growing concern over the possible effects of oil pollution on marine populations. It was thought that molecules from petroleum products could interfere with both smell and taste, upsetting the ability of organisms to feed, socialize, and reproduce.

Jelle moved to Cape Cod in February of 1970. The first question was what organism to study. He needed a creature that was easy to acquire, handle, and manipulate and that had accessible organs of smell and taste. An animal to which electronic devices could be attached would be useful. And Jelle himself had a discriminating sense of taste. A high level of postexperimental edibility would be a bonus.

Jelle knew nothing about lobsters, but a bit of lore he'd picked up intrigued him. Old-time fishermen claimed that a brick soaked in kerosene would attract lobsters to a trap. What better way to begin a study on the effects of petroleum products on undersea organisms than to test this theory? Hoping that his investment would pay off, Jelle purchased a supply of lobsters from the local fishermen's wharf.

Soon Jelle had confirmed that lobsters were attracted to the hydrocarbons in petroleum products such as kerosene. But that discovery would tell him little without an understanding of how odors and tastes governed the behavior of lobsters. Searching for food, socializing, and selecting mates were obvious aspects of lobster life that might rely on water-borne chemicals. Mate selection in particular, Jelle thought, could be revealing.

Smell is an important part of sex for any animal that releases pheromones, including humans. But little was known about how sex pheromones worked in creatures such as insects or crustaceans. All Jelle had to go on was a German study of silkworm moths. The female moth sits in a tree and emits her scent. The male moths catch the scent and fly upwind in a race to reach her. Jelle guessed that lobster mate selection might work similarly. The female lobster would find a place to perch and emit her scent, and the males would sniff her out and come running. The first one to reach her would win.

But Jelle also guessed that some sort of special sex pheromone, and not just any female scent, might be required to induce lobsters to mate. Most of the time, male and female lobsters couldn't stand each other. Prior to the molting season Jelle had paired a male and female lobster in a tank. Immediately the animals had grown agitated and flicked their small antennae rapidly. The small antennae, called antennules to differentiate them from the large antennae, serve as an underwater nose; flicking them is the lobster's way of sniffing.

Detecting each other's scents, the male and female both opened their claws wide and raised them overhead, then whipped the water with their large antennae in a tactile search for the offending odor. The belligerent shoving and snapping of claws that followed made Jelle glad he wasn't in the tank along with his subjects.

Before attempting to mate any of his lobsters outright, Jelle devised a preliminary test. Since females usually mated immediately after molting, perhaps they released a special sex pheromone when they shed their shells. When Jelle's first female molted, he bailed some of the water from her tank into a tank occupied by a male. From the presence of the water alone the male grew agitated, but in a manner quite distinct from what Jelle had observed in his earlier experiment. This time the male sniffed with his antennules, but instead of opening his claws and raising them overhead he closed and lowered them. Instead of adopting an aggressive stance he stood delicately on tiptoes. Instead of hunting cautiously he probed the tank with abandon.

Sadly, he would never find the female that had aroused his libido. Her scent lingered maddeningly in his tank, but by then Jelle had lifted her tender, defenseless body into the tank of another male.

~

Observers of the natural world as far back as Aristotle have wondered how lobster sex works. In the fourteenth century the Italian philosopher and physician Simone Porzio wrote that the lobster's "organs of sex and reproduction are constructed in such a way that I cannot discover any obvious way in which the seed of the male could be ejaculated, poured, or otherwise introduced into the body of the female." The problem was that the male lobster appeared not to have a penis.

Later investigators thought they had discovered the secret

of this vexing omission in the male lobster's swimmerets. Male and female lobsters both possess these little fins, arranged in five pairs along the underside of the tail. But the male's first pair, located at the midriff, are quite unlike those of the female. Instead of the flexible, flattened flippers that adorn the rest of his tail, the male's first pair are hard and pointy. Perhaps it wasn't that the male lobster was missing a penis, but that he had two.

In the 1830s the French naturalist Henri Milne-Edwards, who gave the American lobster its scientific name of *Homarus americanus,* put an end to this speculation. The male lobster's two members, Milne-Edwards wrote, couldn't possibly penetrate the female, on account of their small size. At best, Milne-Edwards thought, the first pair of swimmerets might be "exciting organs"—tools of foreplay.

But leaving female lobsters frustrated wouldn't do, so Milne-Edwards hypothesized the existence of a huge penis that the male kept hidden away except during intercourse. Similar to a collapsible telescope, this membranous appendage would emerge on demand, formed by an erection of the walls of the seminal tube inside the male's abdomen. But the collapsible penis turned out to be a fantasy. By the 1890s the world's first great lobster scientist had set things straight.

Francis Herrick, a graduate student in the zoology department at Johns Hopkins University, was studying snapping shrimp when he chanced upon the lobster for comparison. Snapping shrimp are essentially miniature lobsters. The snap in their name originates from their claws—they stun their prey with the sound of an air bubble popping between their snapping pincers, which clap together so quickly that they emit a flash of light. The American lobster is slower on the draw and can't dazzle anyone with fireworks, but Herrick would come to believe that it did have two penises after all.

The question that most intrigued Herrick about the lobster's sex life was whether the female fertilized her eggs internally or externally. In 1850 a French naturalist named G. L.

Duvernoy had speculated that lobsters never copulated; instead, the male simply fertilized the eggs externally when the female extruded them from her body onto her tail. Herrick didn't believe Duvernoy's hypothesis was correct, but he felt the Frenchman was onto something. Where previous scientists had noted the absence of a lobster penis, Herrick had noted something else—the female lobster's lack of a vagina.

Without a vagina, internal fertilization was impossible. Herrick turned his attention to another part of the female's anatomy. At the base of the female's tail, near the same place that the male lobster had his two hardened appendages, she had a tiny pouch. Herrick realized it was a seminal receptacle, where the female could store a male lobster's sperm until she was ready to extrude her eggs. This gave the female lobster the advantages of both internal and external fertilization. Instead of having to entice a desirable male at the exact moment she extruded her eggs, as Duvernoy had proposed, she could adopt a more opportunistic approach to copulation, mating whenever a desirable male was available. But she also avoided the burdens of pregnancy. She simply kept the sperm on hand in a kind of fanny pack. When she was ready to squirt the eggs out onto the underside of her tail, she unzipped the pouch and performed the external fertilization herself.

Given the position of the pouch, it seemed obvious to Herrick that the male's pair of hardened swimmerets were involved in the transfer of sperm. The appendages didn't need to be long for deep penetration, because the female's receptacle was on the outside of her body. But the skeptics had been correct to point out that the swimmerets weren't exactly penises. They didn't contain hollow tubes necessary for delivering sperm.

But a groove did run along the inside edge of each swimmeret, and their pointed tips matched the size of the hole in the female's receptacle. By dissecting lobster testes, Herrick discovered that the male lobster wrapped his sperm up into gelatinous, tubular capsules called spermatophores. He noted that the duct delivering sperm from the testes opened exactly at the base of the hardened swimmerets—"as if," he wrote, "they served for conducting the spermatophores through the elastic, slit-like orifice of the seminal receptacle."

In essence Herrick had gotten it right. Later researchers confirmed that the male brings his hardened swimmerets together to form rails like a train track, the pointy tips propping open the female's receptacle. Sperm packets are ejaculated through the seminal duct and guided down the rails into the pouch. The trailing edge of the final spermatophore hardens to form a plug that blocks the opening in the seminal receptacle, preventing another male's sperm packets from entering. It isn't exactly a penis-and-vagina situation, but it's pretty close.

Herrick spent five years writing a 250-page treatise titled *The American Lobster: Its Habits and Development*. Published in 1895, it was the first in-depth study of the species. In the text he seems to lament the fact that the lobster misses out on the pleasures of internal copulation. For comparison Herrick describes the mating of crabs, where the male crab runs to the female, embraces her, pulls her to his belly, and ceremoniously penetrates her with his penis.

Herrick needn't have worried. He did his best work on lobsters at a seaside laboratory in the little town of Woods Hole, Massachusetts. Many decades later, working in the same location, Jelle Atema discovered that the American lobster could be a most tender lover, penis or no.

When Jelle's freshly molted female flopped into the new tank, she was so soft she couldn't stand. She didn't need to, however, because the male came to her.

Jelle was worried. This female was his first defenseless subject; having just shed her shell, she was without protection. Jelle hoped the male wasn't simply going to brutalize her, as he might have normally. Jelle was betting that a special sex pheromone, the presence of which he'd still only hypothesized, would do the trick.

Again, the male responded immediately to the soft female's scent. Sniffing with his antennules, he closed and lowered his claws, stood on tiptoe, approached the female, and circled her. As he circled he began to stroke the female's soft body with gentle sweeps of his large antennae.

Not only was the male not going to brutalize her, he evidently had an enormous reserve of patience, for his circling and stroking continued for another quarter of an hour. Finally the female raised herself to a standing position. This indicated, Jelle guessed, that her shell had hardened just enough for the male to handle her safely, and sure enough, the male now mounted the female from behind. Pressing his tail flippers and claws to the floor of the tank for support, he grasped her body underneath his with his walking legs and rotated her onto her back. The two lobsters lay face-to-face, as it were, with the male on top. Both lobsters fanned their soft swimmerets against each other in a flurry of excited stroking. Jelle caught a glimpse of the male's hardened swimmerets pressed against the female's abdomen and saw him thrust several times. Apparently, lobster sex occurred in the missionary position—but with double the male genitalia.

The male dismounted and the female flapped her tail to right herself, then backed into a corner of the tank to rest. The male backed into another corner and adopted a similar attitude of repose. Jelle looked at his stopwatch. After fifteen minutes of foreplay, the act of copulation had taken eight seconds.

Jelle concluded that a molting female lobster must indeed emit some sort of chemical signal that alters the behavior of male lobsters. When Jelle had lowered the female into the tank, the male had been transformed from a belligerent bully into a solicitous master of the boudoir. Whatever constituted this perfume of love, it was a powerful drug.

Jelle repeated the experiment with other lobsters. Sometimes the recently molted female wasn't in the mood and resisted the male's advances. Undeterred, the male would continue to circle and stroke the female's body with his antennae. Jelle termed it a courtship dance, and if the male kept it up long enough he usually managed to persuade the female to let him mount her.

The opposite situation occurred too. Even when the male wasn't interested, the female might sidle up and attempt to slip under him. The tactic didn't usually succeed, but the males refrained from any violence. Something about the presence of the molted female inhibited them from fighting even when it didn't excite them into lovemaking.

Jelle laced an empty tank with female molt water and then dropped in a hard-shelled male and a hard-shelled female, neither of whom had any romantic interest in the other. The two lobsters began an angry quarrel. Just when Jelle was afraid one of them was going to lose a limb, the drug from the absent female began to take effect. The lashing of the male's antennae calmed to a gentle stroking, and the female allowed him to fondle her. Within a few minutes, the goal of beating the female up had become less interesting to the male than the prospect of mounting her, which he attempted several times. The female offered little resistance. She wasn't ready to mate, and nothing came of the male's advances. But the experiment demonstrated how the smell of a molting female alone could modify lobster behavior.

Jelle repeated the experiment, this time with two hard-shelled males. But even a strong dose of female molt water wasn't enough to evince signs of homosexual lobster love. Indeed, with the scent of a willing female wafting between them, the males were more antagonistic than ever.

That suggested a larger pattern. No one had ever seen lobsters mate in the wild, but Jelle imagined how it might work. A female lobster about to molt would seek protective shelter—under a ledge, in a crevice, between two boulders. She would huddle in her hiding place and shed her shell, emitting her sex pheromone into the flowing currents of the sea. A bevy of excited males downstream would catch the scent and scramble upstream to her den, dueling each other for the privilege of courting her. The winner would enter, initiate foreplay, and be rewarded with the opportunity to deposit his sperm packets in her pouch.

Jelle was delighted. His theory of lobster mating matched the German model for silkworm moths perfectly. There was only one problem. His theory would turn out to be almost completely wrong.

4

The Man Show

Ann Fernald was taking a casserole out of the oven when a crowd of men tramped in through the front door. A rush of frigid air followed them.

"Hi, boys!" Ann called from the kitchen. She heard her husband's voice from the den.

"Come on in and take your boots off," Warren said. "And shut that damn door." The men laughed.

Warren had added both leaves to the dining-room table. Ann carried two more serving dishes from the kitchen and worried that the table wouldn't be long enough. She had places for Bruce, Mark, and Dan, along with Jack Merrill and several of the other young men who'd come to the island to go lobstering, as well as her three younger children—fourteen settings in all. That would be enough plates, but would there be enough to eat? She'd peeled an entire sack of potatoes. Warren and Ann had made a habit of inviting Little Cranberry Island's cohort of young lobstermen over for dinner, but Ann was still surprised at how much food they could consume after a day of hauling traps.

Jack came into the kitchen.

"Hey there, Ann," Jack said, giving her a kiss. She patted his ruddy cheek with her stove mitt.

"Can I give you a hand?" Jack asked.

"I think everything's just about set," Ann said. She glanced back at the counter. "Oh, Jack, be a dear and bring in the peas."

Twenty years earlier Ann had come to Little Cranberry Island kicking and screaming. She'd married Warren at eighteen, very much in love, but when he brought her to his island she was miserable. He provided her with food and shelter, but there wasn't much other than winter to be sheltered from—many of Warren's generation had left Little Cranberry, and the place felt nearly deserted.

When Warren started drinking, things got worse. Ann had already given birth to two sons who needed a father when, on the day the third son was born, Warren was so drunk he fell out of *Pa's Pride* into the harbor. By the time Ann had four children Warren was leaving the house some mornings with a lunch pail in one hand and a six-pack of beer in the other. Finally the alcohol had so depleted the vitamins in his body that he had an anxiety attack at the dinner table. He ran from the house and locked himself in his workshop. He had never built a lobster trap so fast in his life.

Ann confronted him that night.

"Enough," she said, pouring a glass of whiskey. She informed him it was the last drink he was ever going to have. Warren had been sober ever since.

Now Ann surveyed the throng of young lobstermen in her dining room and smiled. After two decades of feeling lost, she'd finally found home. And who knew, things might turn around for the island. Two or three of Warren's contemporaries had stayed on Little Cranberry and had sons who were lobstering. Three of Warren and Ann's four sons had become fishermen. The confines of the island could still feel like a curse—one of the Fernald daughters packed her suitcase at the age of fourteen and fled the place, and part of Ann couldn't blame her. But between Ann's sons and their friends who'd come from other places, Ann believed Little Cranberry could be repopulated.

Now, Ann thought to herself, all we need are some young women.

The friendly competition among the young lobstermen didn't stop at catching lobsters. By the mid-1970s Little

Cranberry's year-round population included twenty or so bachelors but only a couple of eligible females. Four generations earlier, three Fernald brothers had migrated to the island in search of women. The Fernald brothers might now have to abandon Little Cranberry to find mates.

Jack had to wonder whether lobstering with the Fernalds wouldn't put an end to his love life altogether. One day he was lobstering with Bruce. Bruce threw a trap overboard and the buoy jammed in the hauler pulley. When it slipped free it shot like a bullet into Jack's groin. Jack crumpled to his knees with tears in his eyes. Bruce was relieved when Jack showed no signs of permanent damage, though of course Jack's recovery did little to ease the competition for females.

When summer arrived on Little Cranberry the warm weather brought a rush of shedders for the lobstermen to catch but also a rush of vacationers to the island. Jack and the Fernald boys worked hard, and they partied hard. A keg on the beach would entice the summer girls from their parents' cottages. In the light of a campfire the young women would contemplate which of the young lobstermen outdid the others in earthy charm and rugged good looks.

Jack was maturing into a thoughtful scholar as well as a handsome fisherman. Between summers on Little Cranberry he'd been attending Antioch College in Ohio, where he'd completed a double major in his two passions, literature and marine biology. He'd supplemented his studies with marine science courses at Boston University, and had even secured academic credit for the lessons in crustacean life cycles he'd learned aboard Warren Fernald's lobster boat. After graduating in 1975, Jack dreamed of becoming a writer, an artist, an adventurer, or all three. He was fixing up an old sailboat for a trip around the world when he realized he needed to make some money. So he moved back to Little Cranberry, bought a used lobster boat, and began building traps.

At a party on the beach one night, Jack got to talking with Barb Shirey, the lithe, dark-haired daughter of a family from Rochester, New York. A year out of college and working in a ski shop near her home, Barb was overflowing with energy, but she was bored.

"Why don't you come live on Little Cranberry?" Jack suggested. "You could be a sternman."

"Who, me?" Barb asked.

"Sure."

"Can you, you know, just do that?"

"I did that," Jack said.

Barb returned to Rochester to think it over, only to learn that her brother had been nursing similar thoughts. So Barb quit her job and returned with him to the island. Her brother secured a job aboard Mark Fernald's boat. But when Barb made inquiries, most of the lobstermen just smiled politely. Jack ought to hire her, Barb figured, but he already had a sternman. Finally Mark agreed to take her, but only because her brother needed a day off. Aboard the boat at four thirty in the morning, Mark grunted and handed Barb a pair of gloves. To prove she was tough, she shook her head.

By midmorning she had realized her mistake. The bones in the fish bait and the sea urchin spines in the traps had lacerated her palms into pink pincushions. With Mark shouting orders at her, she pulled a poker face and worked the rest of the day in excruciating pain. When she got home she sobbed. Her hands were so raw and swollen she couldn't turn on the faucet to wash them.

As autumn settled over the island the nights lengthened and grew colder. In the blackness before dawn the lobstermen would bundle themselves in sweaters and vests, and when the sun rose, an insistent wind would race over the ocean and scratch the tidal currents up into jagged peaks. But the young men had to haul their traps, and had to move into deeper and stormier waters, because the fall was when they made half their money for the year. Lobsters that had summered in depths of eighty or ninety feet continued to shed their shells later in the season because the colder water there delayed the molt. These lobsters were even hungrier than the inshore shedders, and could be caught along with the migrating lobsters that thronged the canyons on their way back out to sea.

Bruce Fernald was planning to fish hard straight through Christmas so he could afford the payments on his new fiberglass boat, the *Stormy Gale*. There was only one problem: his sternman, a strapping fellow with a red beard worthy of a pirate, kept puking. The pounding from the angry waves, the horizon's constant sway, and the flying spray, ten hours a day, wore the man down and turned him green. Bruce couldn't blame him, really. But paying a sternman to be seasick wasn't smart. So he invited Barb aboard.

This time, when offered a pair of gloves, Barb accepted. She rolled up her sleeves and jabbed her hands into the tub of slimy bait. She made a show of stuffing the chopped herring into bags as fast as she could. She manhandled Bruce's heavy traps and rode the pounding sea without complaint. At the end of the day Bruce paid her and asked her to fish with him the next day too. But he didn't say what Barb most needed to hear—that he would hire her for the rest of the season.

Back in Woods Hole, Jelle Atema had built new lobster tanks, and they were huge. Each of the two rectangular enclosures was twenty feet long and held fifteen hundred gallons of flowing seawater. Jelle and his research assistants had spread a layer of gravel and sand across the bottom and piled a cluster of algae-covered rocks in the middle. Along the back wall of each tank they installed cinder blocks to give the lobsters nooks for hiding. The front walls of the tanks were glass, and against these the scientists placed custom-built shelters—a big one at one end of each tank and a smaller one at the other. Jelle and his team had hand-formed the shelters from concrete. They were domed, with an entrance at either end, and they were spacious enough to make a cozy home for a lobster couple. Occupants would find privacy from their fellow inhabitants of the tank, but their activities would occur in full view of the scientists.

To test his theory of lobster mating, Jelle had created as natural an environment as he could inside the lab. In his mind's eye he had envisioned females occupying the shelters, molting and releasing their sex pheromones, and males fighting each other for the privilege of joining a female inside. Catching them in the act would be the hard part. In the wild, lobsters are generally nocturnal, emerging from their shelters after sundown. Yet molting usually occurs during the day, when the lobsters are safely in hiding. If Jelle was to observe social interaction and mating—the latter ought to coincide with molting—he and his team would have to be on call constantly. Jelle and his four assistants drew up a rotating schedule for the late spring, summer, and early fall. Someone was usually in the lab during the day anyway. But every night, seven days a week for six months, one of them would have to stake out the dark lab.

Jelle selected sixteen specimens—four males and four females for each tank. The researchers slipped identification bands onto the lobsters' claws, behind the pincer so the claw remained free to move, before dropping the animals into the water. For realism they added rock crabs, hermit crabs, minnows, and mussels, some of which would serve as food. To help the lobsters adjust to their new environment, Jelle established a routine of switching off the lab lights when the sun went down, and turning on a set of red darkroom bulbs. The lobsters grew comfortable in the tank, and in the red gloom they investigated each other.

The encounters weren't pleasant. In each tank one male quickly established a reputation as a despot. This dominant male bullied the other lobsters until they retreated to a distant corner. In both tanks the dominant lobster claimed the larger of the two molded shelters, leaving the weaker males and females to fend for themselves in the smaller shelters and exposed cinder blocks.

Nothing changed for several weeks. Every night after the lights went off, the alpha male made his rounds, bullying each lobster before moving on to the next. When all the lobsters, male and female, had been dealt their daily humiliation, the dominant male gathered food and returned to the larger shelter.

In a way Jelle could almost sympathize with the beta lobsters in his tanks. A few years earlier, the Woods Hole Oceanographic Institution had refocused its gaze from coastal research to deep-sea oceanography. Research priorities had been reshuffled, and the sex lives of lobsters had come up short. In 1974 Jelle had taken a new job across the street, as an associate professor in Boston University's Marine Program. The BU program was housed inside a respected research center in Woods Hole called the Marine Biological Laboratory. Jelle's new lab was a concrete basement with no windows and only one entrance, but it was spacious and had a steady supply of seawater.

After a few weeks Jelle noticed a subtle change in the social structure in the tanks. The female lobsters had taken up residence in the cinder blocks closest to the dominant lobster's concrete shelter. This was about the time, Jelle calculated, that the females would be preparing to molt. A few days after Jelle made this calculation something else happened. One of the females began to call at the despot's door.

After the dominant male had made his rounds, one of the females he'd abused followed him home and stood by the entrance to his shelter. Visibly agitated, the male turned to face her but stayed inside, flicking his antennules. The female poked her claws through the entrance and flicked back. The male stood on tiptoe and vigorously fanned the swimmerets underneath his tail. The female jabbed the tips of her claws into the gravel, right-left, right-left, and shoved a few pebbles around on his doorstep with her front legs. Then she punched her claws in the male's direction, turned, and walked away.

The visits continued for several days with similar behavior, until one night the female didn't stop at the entrance. The male blocked her way and boxed at her claws with his. But she absorbed the hits and pushed ahead until she was inside the shelter. Then she lowered her claws and turned her tail toward the male, a posture that appeared to placate him. The two lobsters sat uncomfortably side by side. Neither, it seemed, was sure what to do next, and a few hours later the female left. When the two lobsters met outside the shelter, the male acted as if he didn't recognize her. He even slapped her around as usual. But when she showed up on his doorstep again, he tolerated the intrusion. The subordinate males at the other end of the tanks got no lady callers at all.

Soon the female moved into the dominant male's shelter and stayed. She grew irritable, pushing gravel around and turning from side to side. Jelle guessed she was suffering from PMS—premolt syndrome—an activity peak just before the shed. The male now spent most of his time at home and neglected his bullying. Doting on the female, he stood on tiptoe, fanned his swimmerets, and swayed from side to side. One morning after the female had been living with the male for about a week, she appeared especially restless. Jelle guessed that she might be ready to molt.

In the preceding days the scientists had coined a nomenclature to describe behaviors they were observing for the first time, including "substrate jab," "dig display," and "entrance ceremony." What the female did next could be termed only one thing: "knighting." The female stepped up to the male and laid her claws on his head. He stood still, poised on tiptoe, fanning his swimmerets madly. She removed her claws and stepped back while he felt her with his antennae. She knighted him again several times. A few minutes later she fell over on her side, unzipped the back of her shell, and began to wiggle.

~

"How about a game of backgammon?" Bruce Fernald asked.

Barb Shirey had been taking an afternoon stroll along the roads of Little Cranberry Island when she passed the house where Bruce lived with his brother Mark. Bruce had poked his head out the door and called out his invitation.

Behind Barb the road sloped down to the town field, the grass dry and golden in the late-afternoon sun. Barb remembered the way it had been on the Fourth of July, green and bustling with the chatter of picnickers and the yells of a softball game. At summer's end the vacationers had boarded up their cottages and departed, and the only sound now was the rhythmic chirp of crickets. Past the edge of the field, the wharves reached into the harbor, and beyond them the lobster boats hung on their moorings. Bruce's sexy jet-black boat, the *Stormy Gale*, lay in the center of the pack. A chill snaked through the air, and Barb shivered.

"You know what?" Barb said. "I'd *love* to play a game of backgammon."

Hunched over the backgammon board in the toasty living room, basking in Bruce's grin and his naughty fisherman's jokes, Barb thought she could get used to winter on Little Cranberry Island. By the third game Barb was feeling very comfortable. That's when Bruce decided to let her in on a secret.

"You know," Bruce said, "upstairs I have black satin sheets on my bed."

At sea, Bruce had matured into a talented fisherman. He caught a lot of lobsters, and one day he'd even caught an eight-point buck, lassoing the deer right from his boat while the animal was swimming. He'd pulled it in with his hydraulic trap hauler. Now, with his characteristic confidence, Bruce had cast his line for Barb. She was mortified.

"Um, no thanks, I'm not interested."

"Aw, hell," Bruce said, leaning back on the sofa and expelling a heavy sigh. "I shouldn't have said that. Sometimes a guy just doesn't know what to say."

Bruce's words came as a relief, and suddenly the room felt cozier. She couldn't explain why, but Barb felt herself begin to melt.

"Well, how about another game?" Barb offered.

~

Meanwhile in Woods Hole, Jelle Atema had been assuming that the strongest males would pursue and select the most attractive females, and the weaker males would find the less attractive females. But his experiments had shown almost the reverse. The strongest male had simply waited at home. A female had come to his shelter and selected him. The weaker males weren't selected at all.

Jelle now realized that his theory of the female sex pheromone had been backward. Indeed, if the lobsters' behavior in the tanks reflected their actions in the wild, Jelle had stumbled onto something new. What little research had been conducted on the subject so far suggested that in crustacean mating, the males usually took the active role in selecting mates, while the females were passive. In the confines of the thirty-gallon tanks Jelle had used initially, male lobsters had approached molting females. But in the more natural setting of the fifteen-hundred-gallon tanks, the females initiated the approach. Jelle had apparently found a species where the females did the choosing.

A female's sex pheromone was still required for mating to occur, as Jelle's earlier experiments in the small tanks at his old lab had shown. In the big tanks, the dominant males had frequently fanned their swimmerets during encounters with their prospective mate. Fanning created a water current through the shelter, drawing the female's scent in through the front entrance and suffusing the shelter with her smell. As far as Jelle could tell, not only was the female doing the choosing, but by exuding her pheromone she was administering a sort of aphrodisiac designed to turn down the volume on the male's aggression and encourage him toward more tender pursuits. That was quite different from Jelle's original assumption—that a female used her perfume as a homing signal to attract aroused males.

A female lobster would gain obvious advantages by drugging a dominant male. She not only ensured robust genes for her offspring but also secured the best available protection for herself during her most vulnerable moment, the shedding of her shell. The male also gained advantages from this arrangement. Waiting for the female to molt required patience, but it also prevented cuckoldry. The external pouch that formed the female's seminal receptacle was part of her shell. If she had copulated previously with another male, that male's sperm would be thrown out when the female shed, as if she'd been wearing a body suit with a built-in diaphragm. By waiting until the female molted, the male ensured that the offspring would be his.

After copulation the female huddled in a corner of the shelter while her new shell hardened. In exchange, she left her old shell as a postcoital snack for the male. He began nibbling it just a few minutes after dismounting—the lobster equivalent, perhaps, of edible underwear. By evening the female had sufficiently recovered to get up and eat some of her old shell herself. But the male, feeling satisfied, was now keen on cleaning up. He gathered the remaining pieces of her discarded exoskeleton and tossed them in a heap outside the door. The other lobsters in the tank sidled up to the pile like nervous looters and ran off with whatever piece of her shell they could carry.

Over the next few days the dominant male began to leave the shelter regularly again, resuming his bullying and foraging for food. At first the female stayed indoors, her shell still thin as paper. When the male brought food back she would occasionally snatch a chunk for herself. The nourishment strengthened her and she tried to exit the shelter, but the male jealously blocked her way and boxed her claws. One night while he was out, she left.

She came back before morning, and they rested side by side during the day. But they went their separate ways again that night. A few days later the female, her shell growing rigid, was resting inside the shelter when the male returned with food. This time when she tried to help herself he snapped at her. Soon afterward she left and never returned.

In both of Jelle's two tanks, the pairing of the dominant couple had lasted all of two weeks. Things returned to normal, the dominant male continuing his bullying and occupying the best shelter by himself. The females loitered nearby and the subordinate males languished in their ghetto at the other end.

But in one of the tanks the drama took a macabre twist. It came time for the dominant male to shed his own shell. He discontinued his bullying and sequestered himself in his shelter to molt. Wiggling out of his spent exoskeleton, he pumped himself up to an even more impressive size. His new shell was so soft he could hardly stand.

The next time the researchers checked the tank they couldn't find him. Bits of his old shell littered the shelter. Closer examination revealed remnants of body parts scattered throughout the tank. The other lobsters sat in their crannies, stone-faced. They had, it appeared, exacted their revenge.

In the other tank, where the dominant male still reigned in his large shelter, something even more surprising happened. A few weeks after the female moved out, the scientists found standing at the dominant's door a new female, fresh and ready to molt.

~

It was the best round of backgammon Barb had ever played. Despite her obvious talents, Bruce kept Barb working aboard the *Stormy Gale* without offering her a job. She hefted gear, baited bags, plugged claws, and stomached the sea as well as any man. She cracked jokes, played tricks, and exuded energy. Sometimes she even gave Bruce a peck on the cheek to break up the workday. No sternman had ever done that to him before.

At the end of another day at sea Bruce set a course for the island. He crossed his legs, leaned sideways against the bulkhead, and draped his hand over the steering wheel in the lobsterman's classic posture of repose. As he squinted over his shoulder at the water ahead, Barb admired his profile, the nose a straight-edged triangle and the chin chiseled. Her eyes fell to his powerful shoulders and then wandered across his chest. But she was annoyed with him, the sexist bastard. He still hadn't agreed to take her on for the rest of the season. She was sure it was because she was a woman.

"Oh, by the way," Bruce said a few minutes later, "you're hired."

Bruce turned and smiled at his new sternman. Barb pumped her fist in the air and grinned, wanting to kiss him and hit him at the same time.

5

Sex, Size, and Videotape

A fresh new female was waiting outside the entrance to Jelle Atema's laboratory in Woods Hole, and her name was Diane. She had just graduated from college and her passions were courtship and lobsters.

Diane Cowan had fallen in love with the American lobster after writing an English paper on the crustacean in ninth grade. Within a year she had learned to scuba dive. During summer nights on Long Island Sound, while her peers partied on the beach, Diane would slip into the black water with her mask and scuba tank and stalk lobsters on the bottom. Often she surfaced clutching a few specimens for the dinner table. Steaming them and picking them apart was both a gastronomic delight and an anatomic adventure. Diane liked eating males more than females because their claws contained more meat. To her friends who preferred tails, she explained that technically the term "tail" refers to a postanal mammalian structure, and pointed out that a lobster's anus is all the way at the end.

"So this isn't a tail," Diane would state, dipping a morsel in melted butter. "It's an abdomen."

When she graduated from high school in 1978, Diane's friends signed her yearbook "To the Lobster Lady." Her first career move was to talk her way aboard a lobster boat in Provincetown at the tip of Cape Cod.

Diane went to college at the State University of New York at Binghamton. The campus had advantages, including a fabulous professor of animal behavior, but also disadvantages, including the lack of a nearby ocean. While Diane jumped through the hoops of undergraduate life, she amassed a personal collection of newspaper and magazine clippings on lobsters, some of which referred to Jelle Atema's research in Woods Hole.

By her senior year Diane was leading lab sections in the comparative anatomy of fishes, amphibians, and reptiles. When she was assigned an independent research project for an animal behavior class, she decided to study courtship communication. Lacking a local supply of lobsters, she instead chose the green anole, a color-changing lizard often called the American chameleon. Like lobsters, male anoles compete for dominance in order to attract females. They bob their heads up and down, do push-ups, puff up their necks, and pose sideways to accentuate their size. The project required Diane to construct her own anole cages from wire and wood, but her lizards were forever escaping. Usually they were returned the next day by classmates working in distant parts of the building.

One day when Diane walked into her professor's office to discuss her lizards, her attention was riveted by a pair of enormous lobster claws on his desk. She pointed.

"Someday *that's* what I'm going to study," she exclaimed.

The professor asked Diane what she knew about lobsters. He wasn't prepared for the volcano of information that erupted. Diane explained that lobsters communicated using chemicals, and that they were probably attracted to each other's pheromones, as well as to the chemicals in petroleum.

"I'm very concerned," Diane concluded, "that lobsters will migrate toward oil spills, thinking they smell like their sex pheromones, and then all the lobsters will die."

All college students, Diane assumed, were already on an obsessive quest similar to her own pursuit of lobsters. But her professor knew otherwise.

"I have a friend," the professor told her, "who studies lobsters down in Woods Hole. You should write to him and go visit his lab. His name is Jelle Atema."

With this introduction Diane sent Jelle a letter. Hoping for a summer job, she mentioned that she happened to be traveling to Woods Hole—this was a bald-faced lie—and wondered if she could tour his facilities. In April Diane and her mother made the trip to Cape Cod. The village of Woods Hole was still quiet before the arrival of summer tourists. The picturesque main street of shingled shops was bisected by a comically small drawbridge, through which lobster boats and sailboats chugged out from Eel Pond into Vineyard Sound. From the center of town, the laboratory buildings of the Woods Hole Oceanographic Institution and the Marine Biological Laboratory radiated in a maze of cement, stone, and red brick. Along the waterfront Diane saw the Oceanographic Institution's enormous ships, bristling with cranes, antennas, and research equipment.

Jelle greeted Diane and proudly showed her his physiology room, with its Petri dishes and shelves of electronics. She tried to hide her disappointment, for the room contained no lobsters. Then Jelle led her into the concrete basement and she saw the twenty-foot tanks. It was love at first sight.

The infatuation was fleeting, for the tanks contained no lobsters. Jelle had no formal mating experiments planned that summer, and he'd already filled his roster of research assistants on other projects.

"I don't even have a desk you could use," Jelle apologized. But seeing the look on Diane's face, he had a second thought. "If you're going to be around this summer anyway, you're welcome to clean up the tanks and play around with them."

Diane couldn't have cared less about a desk. All she wanted was a chair—and a tank full of lobsters to watch. Diane arranged a babysitting job in Woods Hole in exchange for room and board for the summer. Her plan was to watch the children during the day and the lobsters in the evenings, when the nocturnal creatures emerged from hiding.

After returning home to gather a suitcase of clothes and her guitar, Diane moved to Woods Hole in the middle of May. Jelle was away, so she polished the glass windows of one of the tanks spotless, and at low tide took a bucket to the beach, where she collected starfish, sea urchins, mussels, small crabs, and moss-covered rocks to populate the tank. She even set minnow traps outdoors to stock it with fish. When Jelle returned, he was impressed. A few days later he handed her a bucket rattling with seven lobsters.

"There's a lot we still don't know," Jelle told Diane. "We don't know how females compete against each other to decide which one gets to mate with the dominant male first. All we know is that in our initial experiments, after one female moved out a new female moved in."

Jelle guessed that skewing the gender ratio in the tank might shed light on what was happening. Of the seven lobsters he had given Diane, five were female. The remaining two were a pair of lucky males.

"Keep an eye out for signs of competition between the females," Jelle suggested.

Before dropping the lobsters in the tank, Diane slipped a band around one claw of each animal and labeled them. She named the females F-1 through F-5 and the males M-1 and M-2, though for the purpose of taking notes she decided to use the nicknames M and MM. She equipped herself with a pen and paper and began watching the lobsters in the evenings after work.

Trouble developed quickly, but at first it wasn't inside the tank. The family that employed Diane as a live-in babysitter began asking her to work nights as well as days. Diane was adamant in her refusal—her nights were for the lobsters. So she left.

Finding a new summer residence in Woods Hole was by then impossible, so Diane tucked her suitcase and guitar in a corner of Jelle's lab and took up a nocturnal existence herself. She slept on the beach during the day, woke in the afternoon and went for a swim, then bused tables at a local restaurant until nightfall, when she would return to the lab. Alone under the red darkroom bulbs in the basement, Diane often watched the lobsters until dawn. At one point, she challenged herself to witness, in a single sitting, an entire twenty-four hours in the life of her lobsters. Her record would be eighteen and a half before she collapsed from exhaustion. It seemed an unusual degree of dedication. But Diane had been waiting for this since ninth grade.

Jelle required only a couple of morning encounters with his bleary-eyed volunteer to recognize that her passion was badly in need of discipline. He didn't discourage her marathon sessions in front of the tank, but he taught her how to collect statistically useful data as well as impressions. Soon Diane had drawn a map of the tank and was recording a daily census of the inhabitants, noting each lobster's location every hour on the hour. When she was asleep on the beach or busing tables, one of Jelle's other assistants would take the readings for her. And twice a day at prescribed times Diane conducted what were called focal observations on each animal, quantifying its behavior using a system of standardized codes.

There was no getting around the raw violence on display. The two males, M and MM, were exactly the same size, which prolonged the contest for dominance. Diane recorded forty-six fights between them. M bested his opponent in forty-two of the encounters and finally MM had no choice but to accept subordinate status.

As in Jelle's previous experiments, a row of cinder blocks along the back wall of the tank provided places to hide. Along the front wall there were now four specially molded concrete shelters, larger than the previous ones. Each had two entrances and had been placed against the glass so Diane could see inside. M took up residence in the shelter at one end of the tank. MM bypassed the two middle shelters and occupied the shelter farthest from M, at the other end. Of the five females, one occupied each of the middle shelters and the rest made do with the cinder blocks at the back.

Every evening M would emerge from his shelter, march to the hiding place of each of the females in turn, and bully her until she fled. For good measure M always stopped by the far end of the tank and kicked MM out of hiding too. Then M would strut back to his shelter, slip inside, face the entrance, and wait.

Within days one of the females began visiting the entrance to M's lair. She courted M until he let her inside for a visit. After several visits she moved in, molted, and mated with M. A few days later she was making brief forays into the tank again, though at first she returned to M's shelter after each trip. It wasn't long before she left M's shelter for good.

Almost as soon as she was gone another female was waiting at M's door, and the routine began all over again. The new female courted M, began to visit him inside his shelter, and then moved in with him. She molted, they copulated, and then she began to come and go. After a while she, too, stopped visiting. A day after her last visit, a third female was waiting on M's doorstep.

Again, the third female courted M and moved in with him after a few days. She molted and they mated. After her shell had hardened up she began to come and go. But she was still staying with M when a fourth female snuck in and visited M in the third female's absence. The next day the third female moved out and the fourth female moved in and shed her shell, enticing M to mate a fourth time.

To Diane it looked as if the female lobsters weren't so much competing as cooperating. Yet she'd observed no indication that the females had established any sort of hierarchy to decide who went first, second, third, or fourth. Diane discussed the behavior with Jelle. Biologists had seen similar mating strategies in birds and mammals—the females taking turns so they could all mate with a particular male. There was a name for it: "serial monogamy." The deliberate staggering of molts by a group of females, however, had never before been observed in crustaceans. The female lobsters appeared to be using serial monogamy to ensure that they all had access to the dominant male. But how, exactly, did they do it?

The question grew more complicated when Jelle and Diane discovered that lobsters occasionally bent the rules of serial monogamy. Both genders improvised alternative strategies to subvert the norms of lobster mating. For example, one female suffering from PMS barged into the dominant male's shelter and took a chunk out of the resident female with her claw.

Another who was eager to mate dealt with her dilemma differently. This female—number six—was a newcomer to the tank. After the first two females had mated with M, Jelle and Diane had decided to replace them with fresh females to see how long M could sustain his heroics. M, who continued his tours of bullying, welcomed the new females to the tank with nightly beatings. Female six seemed to find M's advances particularly arousing. But M was already busy with two other females, as his third and fourth mates played hide-and-seek in his shelter. According to the rules of serial monogamy, female six would have to wait her turn. But she was ready to get undressed.

During M's sexual adventures MM had waited forlornly in his shelter at the other end of the tank. The introduction of fresh females turned out to be his lucky break. Female six, fed up with the maneuvering at M's end, paid a visit on MM. After a few days of courtship she moved in. Several times M stopped by and tried to interfere, but female six soon shed her shell, lay on her back, and accepted MM's thrusts.

Males could be equally flexible. For instance, MM spent less time and energy copulating than M did, which gave the subordinate male leeway to pursue other activities—like improving his physique. Perhaps emboldened by his luck with female six, about a week after she moved out MM shed his own shell and molted up to a bigger size. There was a disadvantage to this. Molting took him out of the running for at least two months, the time required for his shell to harden enough for fighting. But it gave MM a 50 percent jump in body weight that he could ultimately use to challenge M, and presumably earn himself some of the new girls. Had Diane continued to add new females to the tank, MM's strategy of outmolting the competition would probably have paid off.

In later experiments, Diane would see another subordinate male molt up and reap the benefits. This subordinate didn't directly challenge the dominant male, but the molt appeared to have given the lobster bigger balls, as it were—he snuck into the dominant's shelter and took advantage of the resident female while the dominant was out.

In another experiment the subordinate fared even better. After he molted up to a larger size, not only had he become the new dominant when fresh females arrived, but he also convinced a brooding female to jettison a perfectly good batch of eggs so she could make new ones with him. This female had mated in the ocean before being captured, and had extruded the eggs onto her tail during her residence in the tank. But impressed with the new physique of the former subordinate, she molted and mated again with him. Once the eggs attached to her shell were discarded—they quickly died—she received sperm from the newly dominant male to fertilize a whole new batch of eggs.

But sadly in MM's case, he had made the wrong move. Diane decided not to add any more females to the tank after he had shed. She wanted to try something else. She wondered what would happen if she skewed the gender ratio in the opposite direction, so that there were more males competing for fewer females.

~

"Speech! Speech!"

The crowd of young lobstermen stood around Bruce Fernald, cheering. A few women were in the room too, including Barb Shirey. Bruce was blushing and trying to contain a broad smile. Jack Merrill stood off to one side. Bruce held up his hands as though he were about to say something, but then Jack walked up and threw a banana cream pie in his face.

The occasion was the founding of the Cranberry Isles Fishermen's Cooperative. Over the previous few months Bruce and another young lobsterman on Little Cranberry had put in hours of research and cajoling, and in 1978 had finally convinced their fellow fishermen to pool resources and buy Lee Ham's lobster dock. Lee, long the island kingpin, was retiring and had put the dock up for sale. By forming a cooperative, Bruce had argued, the new generation of island lobstermen wouldn't have to surrender a cut of their income to a dealer. Using collective bargaining power, they would secure higher prices for their catch, and they would share the profits.

Now the co-op had been operating successfully for several months. Bruce's brothers and fellow fishermen had convened a dinner at the restaurant dock to celebrate. The pie was Jack's way of expressing his appreciation, but without letting things get too friendly. Jack was glad the group would be cooperating for mutual benefit, but half the fun of lobstering was the competition—for lobsters, and for everything else.

To keep up with the Fernald brothers, Jack thought he might need a bigger boat. The investment Dan Fernald had made in his fiberglass lobster boat, the *Wind Song,* had paid off. The hull required almost no maintenance, which meant Dan could spend more time in other pursuits, like fishing for honey holes. The catch had included a daughter of the island named Katy Morse. Seeing Dan's progress, Jack decided to order a fiberglass forty-footer. The only other Little Cranberry man who'd had a boat that long was Lee Ham.

Bruce's fiberglass boat, the *Stormy Gale,* had also proven to be a smart investment, not least because the sexy black hull had helped to entice Barb aboard. Before they knew it, Bruce and Barb were in their third season of lobstering together and Barb had become a dyed-in-the-wool fisherman. One morning the *Stormy Gale* passed Jack's boat at sea. Barb shouted a greeting, pulled down her rubber overalls, and mooned Jack. Another day the *Stormy Gale* passed close by the boat of another young fisherman, who turned, pulled down his pants, and showed his rear. Bruce scooped up a chunk of herring entrails and threw a fastball. The fish parts splattered across the man's ass. He yelped and jerked forward, bumping his groin into the boat's hot exhaust stack.

In winter, the wind stirred up waves so big that slabs of seawater splattered across the *Stormy Gale*'s roof, shuttering the windows and turning the light inside the cabin green. Pouring to the deck, the water would slosh across the floor before draining out the scupper holes in the stern. Barb made a game of the ocean's pounding. She would stand aft of the cabin, and at the moment the boat launched off the crest of a wave she would flex her knees and jump. As the boat sank into the trough, the deck would fall out from under her and she would be suspended in midair, a fisherman flying.

Though neither would admit it, Bruce and Barb both began to nurse a secret thought. If their relationship could survive bitter winds and crashing walls of spray, the storms of marriage ought to be a breeze. But for Barb it wasn't just a question of choosing Bruce. She would also be choosing life on an isolated island. She would be choosing a husband who spent his days riding cruel waves and who came home smelling like putrid fish.

When Barb's third season aboard the *Stormy Gale* drew to a close, she decided she might need a break from lobstering, and maybe a break from Little Cranberry. She and Bruce were on a dinner date at the Holiday Inn, on the mainland, when she told him she was thinking about looking for another job. Maybe a job that wasn't on the island. Bruce put down his fork.

"Hold on," Bruce said. "I ain't having no long-distance relationship."

"Long-distance?" Barb said. "It's only three miles."

"That three miles across the water might as well be a hundred miles."

"Well," Barb said, wondering what she was going to say next, "the only way I'd stay here is if we got married."

Bruce opened his mouth, closed it, and opened it again. "Well, I ain't gonna get married without having my own house to live in." He leaned back in his chair and folded his arms across his chest. He had purchased a piece of land on the island, but at the moment there was nothing on it.

"Well," Barb said, "I'm certainly not going to move in with you anywhere unless we get married."

By the time dessert arrived they'd settled the matter. They'd get engaged if the bank approved a construction loan.

In Jelle Atema's basement lab in Woods Hole, the rivalry inside the tank was intense. Diane Cowan had skewed the gender ratio to four male lobsters and just two females. The competition between males was so brutal that she removed one to ease the competitive pressure. The remaining three continued to clash nightly, and all lost appendages. But they were nearly identical in size and none emerged as the victor.

Without a clearly dominant male, the two females turned fickle and promiscuous, sometimes visiting the shelters of all three males in one night. One of the females eventually settled in with one of the males, molted, and mated. But the other female put off molting altogether, unconvinced that any of the males were worthy of her. Her reluctance didn't dissuade one of the bachelors, who smelled sex in the water. On several occasions he dragged the unwilling female into his shelter, perhaps mistaking her for the other female, who was emitting molt scents from his rival's shelter. He repeatedly tried to mount her and turn her on her back, but her hard shell enabled her to defend herself, and each time she escaped. Meanwhile, without a clear strongman in the tank, the males continued their chaotic fighting. By the end of the experiment one of them was dragging himself around with his mouthparts because all his claws and legs were gone.

The female that had refrained from shedding and mating prompted Diane to conduct another study. She cleared the tank and put in five females—no males. She repeated the experiment four times with different lobsters. Out of a total of twenty females only four shed their shells. In the absence of males, the other sixteen didn't molt at all.

Given what she was seeing, Diane wondered if the assumptions she and Jelle had been making about the importance of the female's sex pheromone weren't too simplistic. Jelle's earlier experiments had corrected the mistaken assumption that male lobsters located an attractive female by her scent, much like the silkworm moth. Rather, female lobsters found the males and then seduced them with their scent. This still assumed, however, that the male's sense of smell was the key to successful sex.

The females were clearly doing some sniffing of their own, however. It was the females that appeared to do the choosing, so perhaps they were targeting a scent released by the dominant male. Even more interesting to Diane was the question of how the females coordinated their molt cycles to achieve serial monogamy.

For example, Diane knew that for female mice, the smell of other mice nearby could influence when a female became ripe for mating. It was the scent of urine that made the difference. A few sniffs of the urine from an adult male would excite a girl mouse, inducing her to reach puberty quickly. If there were too many females nearby that had already reached puberty, the smell of their urine would discourage the girl mouse from taking the leap until her odds were better. In the laboratory, scientists had been able to time the onset of puberty to the hour by mixing an unequal cocktail of adult male and female urine and giving it to a girl mouse to sniff. In human terms the technique could seem a bit crude—just imagine a mother in a pissing contest with the lecherous man next door to decide her daughter's fate. But for achieving the efficient use of reproductive resources, it worked.

Diane guessed that something similar might be going on with her lobsters. The scent of a dominant male would attract a female to the male's shelter. If there was already a molting female inside, her scent would be added to the mix, discouraging the newcomer from shedding her own shell right away. Presumably, the correct cocktail of smells could keep a ripe outsider on the verge of molting for a couple of weeks—long enough for the resident female to do her thing and then clear out. Once the male's scent was ascendant again, the new female would know she was free to move in, molt, and mate. It made sense, especially since Diane had seen the females stopping by the dominant male's shelter nearly every day to sniff for an olfactory update. Diane knew that similar olfactory cues had been shown to synchronize the menstrual cycles of human females living together in college dormitories.

Diane hypothesized that the female's sense of smell might be just as important to successful sex as the male's. There was only one way to find out, of course. Cut off their noses.

Inside Jelle's lab Diane had been promoted to Ph.D. candidate, and outside the lab she had secured an agreeable place to live. She'd come a long way from days sleeping on the beach and nights sitting in the concrete basement. As much as she loved lobsters, she liked spending some of her nights at home. A new video recording system in the lab allowed her that luxury. She would continue to observe social life in the tank firsthand at regular intervals, but the video cameras would give her the chance to enjoy a social life herself.

The experiments would be the first Diane had designed on her own. The idea was to repeat the earlier mating scenarios with four females and two males in each tank, but deny some of the lobsters the ability to smell. She started with the males. With a pair of scissors she snipped their antennules off before plopping them in. Four females went into each tank untouched.

Without their antennules, Diane's male lobsters wouldn't be able to smell, but they would still be able to feel their way around the tank using their long antennae and the hundreds of feeler hairs on their claws, legs, and body. The lobsters would still be able to eat when they stumbled—literally—onto food. They had taste receptors on their feet and mouthparts. But if mating depended on male lobsters being able to smell the female sex pheromone, these males might as well have been castrated.

When Diane reviewed the videotapes she saw an immediate difference. Without noses the males didn't fight. In fact, they paid no attention to each other. As a result, the females in the tank had no clue as to which male was dominant, and most of the females elected neither to molt nor to mate. However, between the two tanks, three of the females still considered the males worthy of seduction.

When these females began calling at the entrances to the males' shelters, the males did not respond by standing on tiptoe or fanning their swimmerets as they should have. Nevertheless, after several more visits the females chose their males and pushed their way in. Unable to smell, the males were belligerent and two of the females were injured in the ensuing spats, but all of them managed to get inside without being killed. Diane began to wonder if this experiment had been a good idea. The females would soon be shedding their shells. If they were counting on the males to protect them instead of eat them, things could get ugly.

Diane was in the lab on the morning when the first female broke her shell membrane and fell over on her side. Her old covering came off and she lay exposed. Instead of standing guard nearby while her new shell congealed, the male approached the female and stood over her. Diane watched as he unfolded his feeding mandibles and tasted the female's soft tissue. Diane expected the worst. Then something funny happened. Instead of tearing off a chunk of her flesh and devouring it, the male just kept tasting.

The tasting continued, and the male climbed on top of the female and flipped her over—sooner than he would have normally. Though she was unusually soft, she endured the male's copulatory thrusts and received his sperm successfully. Astonished, Diane could only conclude that rather than smelling the female's love drug, the male had tasted the sex pheromone on her body instead—he'd fallen into a romantic stupor induced not by sniffing, but by licking. The other two females also mated successfully with their denosed males, who were clumsy but not violent. Apparently, even without their olfactory organs male lobsters could still be induced to sex through oral stimulation.

Curious as to whether the reverse would be true, Diane reset the experiment and dropped two fresh males in each tank, antennules intact. Then she snipped the antennules off ten females and dropped five in each tank. When she watched the videotapes the drama commenced normally. The two males dueled to establish dominance. The winner made his rounds, beating up the females and strutting back to his shelter, where he waited for the stream of lady callers. But they never came.

The females wandered the tank aimlessly, accepting their daily beatings without rewarding the dominant male with even the most basic pleasantries of courtship. Two of the females eventually made overtures, but their advances were brusque and ill-mannered. Normally courtship and cohabitation lasted a couple of weeks. These females pushed their way into the shelter, shed, copulated, and moved out in two days. It was the lobster equivalent of a one-night stand.

Four of the other females, unable to assess their social situation by smell, molted carelessly and without male protection, leading to humiliations that made Diane cringe. When it came time to shed, all four of them lay down in the middle of the tank and exposed themselves. For one of them the results were catastrophic. In effect, she was raped, then killed.

Diane's video cameras were outfitted with silicone intensifier tubes for night vision, but the recorded images of these particular crimes were dim and grainy. The unsolicited copulation had obviously been inflicted by a male, but the killing could have been anyone's handiwork—Diane had seen hard-shelled females butcher soft females before. She replayed the tapes but couldn't identify the shadowy attackers.

It turned out that the death of the female wasn't the only drama Diane missed. She later noticed that four of the females that had exposed themselves in public had been badly injured by mysterious assailants. But it got worse. After the experiment Diane dissected the females that had molted and discovered several whose seminal receptacles contained sperm. They had been beaten and raped too. It was tempting to blame the males in the tank, but Diane hadn't treated the females much better herself. She'd snipped off their noses, used them as pawns in the game of science, and then sliced them open to probe at their privates. It was a nasty business all around.

Cutting the antennules off males had left them pugnacious and inept, but the females had still managed to cajole the noseless males into a standard courtship routine. Cutting the antennules off females, by contrast, had nullified the routine and caused chaos. To Diane the experiments suggested that males were secondary actors in an olfactory drama primarily of females, whose ability to sniff their way to successful sex was the key to mating. It was a skill that sustained a kind of lobster sisterhood, where olfactory cues allowed females not only to identify the dominant male but to schedule their moments of unarmored availability to take advantage of his presence. When the system didn't work, the sisterhood suffered, leaving female lobsters isolated and vulnerable to undesirable males, and probably to each other. Perhaps it was just as well that human females in college dormitories hadn't developed a similar system.

~

Flashing a smile, Bruce Fernald hefted the beer can with all the machismo he could muster and tugged the tab, spraying a burst of alcoholic foam at the camera. The attractive women that surrounded him on the beach clutched at their own cans of beer and cheered. The director flapped his script at the cameras and called it a wrap.

The outside world had come to Bruce's corner of Maine to honor the life that he and the other fishermen of Little Cranberry Island had chosen. Bruce's days at sea were an unfolding drama of man against nature, embodying a frontier spirit that struck a deeply American chord. What better way to honor the American lobsterman, Bruce thought, than by surrounding himself with beautiful ladies while filming an Old Milwaukee beer commercial?

Of the local lobstermen who auditioned, Bruce was deemed one of the most photogenic. The director chose him to drive the boat, to heft the big lobster on the platter, and to sit with three buddies as they raised their cans of Old Milwaukee in salute to the lobstering life. When the ad aired, Bruce loved seeing it on television. There he was at the end, agreeing that "Boys, it doesn't get any better than this."

In fact, it was about to get a lot worse. While Bruce was busy filming a beer commercial, the outside world had arrived in Maine in another, less welcome way. In 1970 President Richard Nixon had created the National Marine Fisheries Service and charged the new agency with extracting the maximum sustainable benefit from the oceans. In 1976 Congress had passed the Fisheries Conservation and Management Act, which made a national priority of defending America's fish stocks from the ravages of overfishing. In Maine, government scientists had embraced this task with fervor, and what they saw in Maine's lobster industry seemed cause for alarm. To them, the man-against-nature drama was probably a bad thing because nature appeared to be losing.

A fisheries expert named Robert Dow had been studying Maine's lobster fishery for years. By the late 1970s he was worried about two disturbing trends. Put simply, lobstermen appeared to be having too much sex, while lobsters weren't having enough. The number of new lobstermen on the coast had nearly doubled, while at the same time water temperatures in the Gulf of Maine had dropped, causing Dow to fear that lobsters were becoming less active. Dow suspected that in colder water lobsters mated less, and fewer of their offspring survived.

A dramatic decline in the lobster population could lie ahead, Dow said, when fishing effort was at an all-time high. Thanks to ambitious young lobstermen like Bruce Fernald, his brothers, and Jack Merrill, the number of traps along the Maine coast had risen to nearly ten times the historical average. Dow's data indicated that catches had already started to fall. Dow implored the industry to cut back before it destroyed the population, but few listened.

The lobsterman, Dow finally stated in 1978, "is a shortsighted, monopolistic exploiter of a public resource." Fishermen, proving themselves to be rapacious and greedy, were failing to protect the lobsters they depended on for sustenance. "If they're stubborn much longer," Dow concluded, "we won't have a resource to worry about."

"All the data indicate that we're in for a steep decline," another scientist in Maine told the press. "The thing's going to crash on us. I feel very bad about this. I know it's going to come."

PART THREE

Fighting

6

Eviction Notice

If Bob Steneck were a lobster and wanted to cover his butt, what sort of shelter would he choose? Something that was easy to back into and defend from attackers. Bob sawed a PVC pipe into foot-long sections and tacked a rubber flap over one end of each piece. Donning his scuba gear, he descended underwater with his tubular homes and arranged them on a barren expanse of sediment in two rows, widely spaced. The rubber flaps were on the outside ends, so the entrances to the pipes faced inward, toward a kind of public square. It looked like a nice neighborhood.

Bob had no business building lobster neighborhoods. He'd been hired at the University of Maine as a marine ecologist in 1981 to study more arcane matters, like how long it took a sea urchin to eat a leaf of kelp. Bob had already made a name for himself piecing together an epic battle between coralline algae and vegetarian snails in the Caribbean, an evolutionary arms race that had transpired over millions of years. When Bob arrived in Maine he set out to examine the feeding patterns of herbivorous echinoderms and gastropod mollusks—the sort of blobs in shells that were known locally as urchins, snails, or limpets—but on his dives he was constantly distracted by lobsters. In the Caribbean Bob had sometimes glimpsed clawless spiny lobsters, but in Maine the lobsters were so plentiful and active that he had a hard time not playing with them. The animals seemed happy to oblige. When Bob checked his newly constructed neighborhood the next day, a pair of claws and antennae were poking out the entrance of every pipe.

"Shit," Bob said to himself, "why the hell am I studying limpets?"

Getting underwater had been an obsession for Bob ever since his boyhood summers at his grandparents' house on Lake Hopatcong in New Jersey. By the age of ten Bob had set a family record by swimming the four miles across the lake and back, but the surface of the water wasn't what interested him. He devised a makeshift scuba tank from a plastic-lined canvas sack. The buoyancy of the air-filled bag prevented him from reaching the bottom, so he tied window-sash weights made of lead around his waist to drag him down. His friend on shore was supposed to replenish the air in the bag through a garden hose attached to a bicycle pump. Bob had been on the bottom watching a crayfish for more than an hour when he got a headache. Surfacing, he found that his friend had abandoned the pump and gone fishing.

Now in Maine, Bob tried observing lobster behavior the same way he'd watched crayfish as a kid. Even with proper scuba gear, however, it was nearly impossible because lobsters could detect the minutest movements of water. If Bob tried to sneak up on a lobster that was foraging or searching for a shelter, the animal sensed the pressure waves emitted by the bubbles from Bob's scuba regulator, stopped what it was doing, and turned to face him, its claws raised.

Bob wasn't the first scientist to encounter this problem. Poring over books and journals in the library, Bob discovered that little was known about the behavior of the American lobster in the wild. But scientists had managed to learn some astonishing things about the way lobsters used habitat. In particular, lobsters boasted an impressive repertoire of excavating and remodeling skills. By spreading its mouthparts and front legs into the shape of a bulldozer blade, a lobster could shove sand or gravel from a burrow to enlarge it and erect barricades against intruders. Lobsters were also efficient stone movers. By carrying pebbles with their mouthparts or rolling larger chunks of rock, they built perimeter fences and blocked off strategic openings in a burrow.

When faced with open terrain lacking natural crevices, a lobster was capable of creating its own burrow, though this was a time-consuming process. Like a dog digging for a bone, the lobster scratched the sand or mud out with its front legs and flung it back between its hind legs, where it piled up under its tail. Between bouts of digging, the lobster fanned its swimmerets to clear excess debris, a cloud of sand billowing out behind it. Lobsters were partial to a home with an escape hatch—an exit in the back that was small enough not to require constant defending, but that allowed the lobster to slip out in an emergency. If the crevice or hollow didn't come equipped with a back door, the lobster would often excavate one.

Bob chanced upon several papers written by Stanley Cobb, a marine biologist at the University of Rhode Island. Stan Cobb's research suggested that the American lobster's defensive strategy was governed partly by the desire to have its body touching things and partly by the desire to avoid light. The result was an affection for the coziest and gloomiest recesses of the seafloor, which provided defense not only against predators and other lobsters but also against the tidal currents that could rip through an underwater channel like blasts of air through a wind tunnel. Hiding under a flap of seaweed would suffice in a pinch, but a low-ceilinged rock hollow was best. Stan's experiments with clear and opaque domes demonstrated that darkness was more important to lobsters than snugness, but a lobster under bright lights would rather back itself into a glass jar than wander in the open.

Neglecting his urchins, snails, and limpets, Bob enlisted the help of several scuba divers and took a census of lobsters on the seafloor between Casco Bay and Penobscot Bay, near his office at the Darling Marine Center, the University of Maine's coastal lab for ocean science in Walpole, Maine. It didn't take long for Bob and his divers to realize that there were good neighborhoods for lobsters, where the population density was high, and bad neighborhoods, where lobsters were scarce. Not surprisingly, the good neighborhoods tended to be boulder fields with lots of nooks and crannies. The bad neighborhoods tended to be featureless bedrock or flat sediment with nowhere for a lobster to hide.

Bob had earned his Ph.D. at Johns Hopkins University in ecology and evolution. He'd learned that an ecologist's first job was to notice a pattern in nature and then simply to observe it for a while. Such patterns usually had to do with the distribution and abundance of organisms—where, and how many? Once Bob had observed a pattern to his satisfaction, his next task was to come up with a hypothesis about the natural process that might have created that pattern. Finally, in testing his hypothesis, he would try to identify a concrete mechanism in nature that was responsible for the process. The hope was thereby to gain some insight into how organisms had evolved, especially in relation to each other.

It was a simple creed—patterns, processes, mechanisms—and examples were everywhere. When Bob, wearing shorts, strolled through a field and stumbled into a patch of thistle, his mind didn't stop at "Ouch." The thorns drew his attention to the pattern of distribution and abundance of different plants in the field. That suggested a process—grazing by animals. The mechanism that drove the process was the mouth of the animal, the tongue and flattened teeth evolved perfectly for munching grass. In turn, the perfect defense against that mouth was to evolve thorns. Thistle could take over a field because it had become more resistant to grazing than other plants. It hadn't taken Bob long to realize that the only constant in his line of work was that populations of organisms were always in flux.

Now Bob had noticed a pattern in the distribution and abundance of lobsters. The seafloor had good lobster neighborhoods and bad ones. Bob wondered if the terrain of the seafloor itself might even control the number of lobsters that could live in a given area. Young lobsters might need swaths of small rocks for hiding, older lobsters bigger boulders.

Lacking either, they would jog off in search of more protective terrain.

Bob sawed up more PVC pipe and built additional lobster neighborhoods on the sediment. He had commandeered a leaky old houseboat from the Darling Marine Center, and now he anchored it above his arrays of pipes. Bob set up a generator on the boat and ran a video cable into the water. On the bottom it was attached to a miniature ROV—a remotely operated vehicle—with tiny propellers, low-intensity lights, and video cameras, none of which bothered the lobsters. Instead of distracting the animals with his scuba bubbles, Bob could sit aboard the houseboat all night long, hunched over the glow from a television set, and fly the ROV over the bottom to watch what the lobsters were doing in the neighborhoods he'd built for them.

~

The new house on Little Cranberry Island was white with blue trim, square and snug, set back from the road at an angle. Bruce Fernald put his arm around Barb as they gazed at it. They could hardly believe it was finished. They'd hired a couple of friends to build most of the house, one of them an island contractor and the other Bruce's former sternman with the predisposition for puking. The man was much happier on a roof than in a boat.

Bruce and Barb had assisted with the construction of the house by pounding nails and painting. Barb planned to plant a garden in front. Bruce had outfitted the back with a basement entryway for carrying his lobster traps, buoys, and coils of rope in and out. The living room was cozy, with a low ceiling for easy heating. In the kitchen, sunlight streamed onto the breakfast table by the window. Next to the sink was a VHF marine radio, and from the roof rose an antenna so Barb, now retired from lobstering, could call Bruce on the boat. The house was nestled in a stand of spruce that served as a buffer against winter winds. The blanket of trees stretched for half a mile down the road to the island's seaward beach, where a barricade of gray cobblestones kept the ocean at bay. Two or three of the island's other young lobstermen had built houses nearby.

The wedding took place in July 1979, on what began as a foggy day. As people gathered on a lawn by the shore facing Mount Desert Island, the fog lifted, leaving a lacework of clouds hung like a bridal veil, curving with the contours of the Mount Desert hills. Bruce, solid and beaming, sailed across the grass in a trim three-piece suit of sky blue. Barb, elegant and giddy, was at his side in an off-white eyelet dress with a camisole top. She clutched a bouquet of island daisies and roses in one hand and Bruce's forearm in the other. Virtually the entire island community encircled them on the lawn. After the ceremony a metal skiff with an outboard motor pulled up to the rocky beach and took Bruce and Barb on a joyride around a neighboring island. July was the beginning of the trapping season, and for now a boat ride was all the honeymoon Bruce and Barb would get.

The wedding was a joyous celebration for the community, but back aboard their fishing boats the young lobstermen of Little Cranberry worried. They knew that seventy-five miles down the coast, scientists at Maine's Department of Marine Resources had been studying the lobster population. The Little Cranberry lobstermen were finding plenty of lobsters in their traps. Nonetheless, the scientists' prognosis for the future was not good.

~

To find out how the lobster population was faring, scientists at Maine's Department of Marine Resources had tagged and released lobsters and then put up "wanted" posters at the docks, asking fishermen to report those animals if they were caught—a clue to the rate at which lobsters ended up in traps. In addition, the scientists hauled some lobster traps of their own and analyzed the catch. They also traveled the coast, recording the size of the lobsters that fishermen sold at the wharf. Back in the lab, they dissected big and small lobsters to determine the size at which a lobster's reproductive organs became functional. And they noted the size of the lobsters that extruded eggs while in captivity, to determine how large a female lobster had to grow before she was capable of bearing young.

The scientists made two crucial findings. First, if their statistics were correct, the lobster industry's annual harvest was composed overwhelmingly of lobsters that had just molted up to the minimum legal size. In Maine the smallest legal lobster was defined as an animal whose "carapace length" was at least three and three-sixteenths inches. In the 1800s lobsters had been measured from the tip of the bony spike between the animal's eyes to the end of its tail flippers, but this definition had proved too easy for fishermen to fudge. In 1907 the method was changed to the length of the carapace, which is the single large section of shell that encompasses the lobster's thorax—often referred to by restaurant-goers as the "body" of the animal, to distinguish it from the meat-filled "tail." To measure the carapace, a lobsterman would line up one end of his "gauge"—the brass ruler he carried aboard his boat—at the lobster's eye socket and measure back to the edge of the carapace. When a lobster had molted to become large enough to pass this minimum-size mark, it weighed, on average, just under a pound—a meal of modest proportions.

The scientists discovered that ninety out of every hundred lobsters that fishermen brought to the dock each year were these modest-sized animals, only just big enough to meet the measure. The other ten lobsters would be just one or two molt increments larger. With lobstermen saturating the seafloor with traps as never before, the scientists believed that few lobsters made it through the gauntlet of fishing gear to grow much bigger than the minimum size.

The second finding had to do with how soon lobsters became sexually mature, and it lent an ominous cast to the first finding. The geographic range of the American lobster extends from North Carolina to northern Labrador, near Greenland. In southern latitudes where the ocean is warm, the reproductive organs of lobsters begin functioning at a smaller body size than in northern latitudes. In Long Island Sound, for instance, at a carapace length of three and three-sixteenths inches nearly all female lobsters have reached puberty and are capable of mating and extruding eggs. In the colder waters of the Gulf of Maine, however, the scientists calculated that only 6 percent of females were sexually mature at Maine's minimum legal size.

Combined, the two findings pointed to a devastating conclusion. Hardly any female lobsters in Maine got the chance to reach puberty—let alone mate and make eggs—before they ended up on someone's dinner plate. With fishing on the increase and catches in decline, the situation seemed grim. In the Gulf of Maine, the American lobster could well be teetering on the verge of a disastrous collapse.

If so, the number of lobster young in the ocean needed a revitalizing boost. By the early 1980s the scientists were pushing a plan to raise the minimum carapace length from three and three-sixteenths inches to three and a half inches. As a result of this larger minimum size, they calculated, the fraction of females capable of making eggs would jump from 6 percent to 60 percent.

Lobstermen couldn't think of a worse idea. Instead of weighing less than a pound, a lobster with a three-and-a-half-inch carapace would weigh at least a pound and a quarter. At the restaurants and retail outlets where summer vacationers purchased their lobsters, a pound and a quarter was the weight that separated small "chicken" lobsters from "selects." Chickens—or chix as lobstermen called them—could be bought by consumers on the cheap and were thought to be popular for clambakes and informal festivities, while selects were pricier. If the scientists' recommendations were followed, the cost of a typical lobster dinner on the Maine coast might rise by several dollars.

With much of Maine's catch consisting of chicken lobsters, lobstermen worried that they could lose the lucrative market for chix to their competitors in Canada. They also worried that implementing the new policy would cost them most of their catch for a couple of years while they waited for the first round of lobsters to grow up to the new minimum size.

At Little Cranberry Island, Bruce Fernald had a construction loan to pay off and Jack Merrill had sunk so much money into his new forty-foot boat that he'd christened her the *Bottom Dollar*. Bruce and Jack feared that if the scientists had their way, hard times could hit the island. More to the point, they didn't think boosting the egg supply was necessary in the first place. They were cutting V-notches in plenty of female lobsters, and they saw those female lobsters repeatedly making eggs. To the young lobstermen the future looked promising. It did to the older lobstermen too.

Warren Fernald headed down the coast to a meeting and learned that lobstermen from other parts of the state were equally upset. Some had coined a name for the lobster scientists. The "bug hunters," they said, were wrong.

At the meeting Warren encountered Maine's commissioner of marine resources. They'd met before, and the commissioner considered Warren one of the friendliest fishermen he knew.

"I challenge you," Warren said, "to come to my island. You're invited to stay at my house overnight, and my wife Ann and I will feed you supper, and you can haul traps with me the next day."

The commissioner took Warren up on his offer. In the morning Warren woke him at five thirty. Aboard the *Mother Ann,* Warren made the commissioner stuff bait bags and empty traps. They spent most of the day throwing lobsters back into the sea. The shorts, eggers, notchers, and oversize lobsters that Warren tossed overboard far outnumbered the lobsters that he plopped into his barrel of keepers.

The commissioner was impressed by Warren's efforts at conservation. He hadn't known that fishermen like Warren threw so many lobsters back.

"How the hell would you?" Warren said, grinning. "Up there in your office, how would you know what was in these traps?"

For Bob Steneck, camping out in the Darling Marine Center's leaky houseboat all night to watch lobsters was nothing new. Bob had been putting his own life on hold to probe the mysteries of nature for as long as he could remember.

As a boy Bob had loved the trips to his grandparents' house on the lake, but he was less fond of the obligatory social calls on family friends and relatives. Instead of visiting indoors he would crawl on his hands and knees into the bushes outside, scanning for the sandy funnels that betrayed the dens of ant lions. Also referred to as doodlebugs, for the meandering trails they left while searching for a location to dig, ant lions waited with terrible clawlike jaws at the bottom of their excavated funnels for hapless insects to fall in. Bob would help secure a supply of doomed ants, but if none were handy, the lion could be tricked into attacking a rolling pebble.

Later Bob joined the junior birding club, and when he moved to Ohio to attend college in 1969, he met an elderly professor of natural history who taught him ornithology. Developing a passion for ducks, Bob mentioned to his professor that he'd love to see some wood ducks. The next morning Bob's professor picked him up before dawn, drove through the darkness, and pulled over on a deserted shoulder. The older man disappeared into the underbrush on his hands and knees. Bob followed. After a long crawl they reached the edge of a swamp. It was deserted and Bob began to fidget, but five minutes later a pair of wood ducks flew in and landed a few feet away.

The passion for ducks passed, but Bob had learned a valuable lesson that morning. If you searched with persistence, and observed with a keen eye, Mother Nature would eventually reveal her secrets. In the Caribbean, Bob had spent an entire year living in the pontoon of a trimaran sailboat, diving daily to study algae. Now he was camping aboard the University of Maine's houseboat to watch lobsters all night, and in the face of his vigilance the lobsters were yielding clues to their behavior. But they weren't the clues Bob had expected.

Bob already knew that lobsters fought with each other over

access to shelters, and that evictions in lobster neighborhoods were common. In Stan Cobb's lab at the University of Rhode Island, one of Stan's graduate students had watched lobsters negotiating for ownership of the more desirable shelters in the tanks. Often the evictions were decorous, even polite—especially when the evicting lobster had an overwhelming size advantage. The larger lobster would approach the entrance and rap lightly on the resident's claw, as though knocking on a door. The bigger animal would then spin sideways and walk backward about a body length, making room for the evictee to emerge. The smaller lobster would exit the home, turn to face the intruder, and retreat. As the small lobster backed up, the large lobster would walk forward. Once the large lobster had passed the entrance to the shelter, it would stop and back in, claiming the home for itself. It was a carefully choreographed dance.

When two lobsters were more evenly matched the resident was less willing to surrender its home. Some intruders pushed their way in through the door and shoved, boxed, or snapped at the resident before retreating to let it vacate. Others dispensed with the preliminary encounter and simply backed in, forcing the tough armor of their tail into the resident's face. This was a tactic shared with the California mantis shrimp. A male mantis shrimp will try to evict a male resident of a burrow by presenting its tail. The resident will strike at it with its claw before turning around to have its own tail struck—essentially, male dominance determined by spanking. New England's lobsters were more reserved. If presented with a tail, the resident lobster would step out of the shelter and turn back to contest the eviction, having now lost the advantage of residency.

Sometimes a large lobster would even annex the shelters of smaller lobsters to form an addition to its own home. In a tank outfitted with a Plexiglas board on top of a pile of sand, Stan's student gave four lobsters the chance to dig their own shelters, visible through the Plexiglas. One lobster was larger than the others, and within days it had kicked out its neighbors and amalgamated their homes into a sand mansion five feet deep and outfitted with four doors. Two of the evictees became homeless refugees; the third squatted in a tiny nook of sand at the other end of the tank.

Given the results of these laboratory experiments, Bob Steneck had assumed that in the ocean, larger lobsters would always evict smaller ones. To some extent this was true, but Bob saw that the preferences of different lobsters came into play as well. While conducting his censuses of lobster neighborhoods, Bob had asked one of his interns, a former violin maker, to construct a device for measuring the interior dimensions of lobster homes. Underwater, when Bob or one of his assistants saw a pair of claws protruding from a crevice, they would coax the creature out, capture it, and record its body size—a dicey proposition if the lobster was large enough. Then they would insert the measuring tool into the vacated shelter, record the length of the hollow, and push a lever that spread out feelers to probe the diameter. It turned out that a lobster would shop around for a home that fit its particular preference relative to its body size. To Bob the behavior wasn't unlike humans picking out blue jeans. He devised a record-keeping system in which young lobsters preferred "restricted-fit" shelters. Older lobsters also tried on restricted-fit shelters, but many seemed to prefer "relaxed-fit."

If a large lobster liked relaxed-fit, it might not bother evicting a smaller lobster from a restricted-fit shelter. In the neighborhoods of pipes Bob built, large lobsters left small lobsters alone in their restricted-fit shelters—but only as long as the neighborhood was zoned to allow each lobster command of a spacious yard. During the night, while the lobsters were out, Bob moved the pipes closer together. When they returned, the bigger lobsters strutted around the public square bullying the little lobsters until they moved out. In most cases, the dominant lobsters didn't want to live in the restricted-fit pipes themselves. They were just annoyed at having inferior neighbors in such close proximity.

When Bob tightened the zoning again, so that the neighborhood was now compressed like a city block, something even more peculiar occurred. When the lobsters returned at dawn, so many small lobsters were vying for the restricted-fit pipes in such close quarters that the big lobsters simply gave up, and this time it was they who moved out. Apparently, constant aggravation was too high a price to pay, and the dominant lobsters left to seek out a less populated neighborhood.

Even for a big lobster, then, avoiding conflict was sometimes the best alternative. Unfortunately, this was a lesson that Bob would fail to learn. For a conflict was brewing between lobster scientists and lobster fishermen, and Bob was about to strut into the middle of it.

On Little Cranberry Island, Katy Morse Fernald removed the plastic lid from one of the empty coffee cans she'd collected, cut a rectangular slot in it, and replaced the top. Then she picked up a wooden lobster-claw plug from her husband Dan's supply and dropped it through the slot. It fell into the can with a satisfying plunk. She took the tops off the rest of the coffee cans and cut a slot in each.

When Katy Morse married Dan Fernald in 1977 she had brought to the union an island pedigree only a little shorter than his. Katy's grandfather had come to Little Cranberry in 1885 as a fifteen-year-old orphan looking for a job skinning fish. Now Katy was hitched to an island fisherman and was planning a family. But if the government scientists were right, and the lobster population was in trouble, she worried that not even Dan's lightning-fast hand at lobstering could help them make ends meet.

In 1954 Katy's father-in-law, Warren Fernald, had been a founding officer of the Maine Lobstermen's Association. The three young Fernald brothers were now members, and Bruce had been elected to the board of directors. Jack Merrill, with his interest in marine biology, was active in the association as well. At the MLA meetings the men grumbled about the bug hunters, their pessimistic predictions, and their plan to impoverish lobstermen. Katy was completing her bachelor's degree in economics at the University of Maine and was writing her thesis on the economics of the lobster fishery. None of the Fernald men had gone to college, so Katy thought she might have something to contribute and started tagging along to the meetings. At first the men turned their grumbling on her, mumbling about how MLA meetings were no place for a fisherman's wife. Then they heard about her plan to help them with the coffee cans.

Katy replaced the lids and distributed the coffee cans around the island. The plan was simple. Her husband, brothers-in-law, and father-in-law would take the cans out on their boats, and every time they tossed a V-notched lobster back into the sea, they would plunk a wooden claw-plug into the coffee can. When the cans were full they'd dump out the plugs and bring them home for Katy to count. By comparing the numbers of claw plugs with the catch records from the co-op, Katy could get a rough estimate of how many protected females around Little Cranberry were producing eggs. If fishing communities like hers could tell the scientists how many V-notched lobsters were in the water, lobstermen might be able to show that the lobster population was already protected.

Katy's plan was well conceived, but the problem was not simply that the government scientists lacked information. They also lacked trust in the lobstermen's claims. The suggestion that fishermen would protect lobster eggs of their own volition was ludicrous to anyone who knew the history of the fishery. Because in truth, lobstermen had a terrible track record.

In the nineteenth century lobster eggs had been a delicacy popularized by the chefs of London's West End, who mashed them into sauces or sprinkled them on salads. Crustacean caviar had less culinary value in America, but the lobsters to which the eggs were attached were a valuable catch. In Maine, state authorities recognized as early as the 1870s that harvesting egg-bearing lobsters was a bad idea and outlawed their sale. But lobstermen just laughed and scraped the eggs off with a stiff brush, slaughtering millions of embryos and removing thousands of mother lobsters from the sea.

In the early 1900s the government changed tactics and instituted a buy-and-release program for egg-bearing lobsters. Lobstermen just laughed louder because the taxpayers of Maine were now paying them to catch the same female lobsters over and over again. Not to be ignored, some clever bureaucrat came up with a new use for a paper hole punch. Punching a hole in a lobster's tail flipper before the animal was released indicated that the lobster was government property and couldn't be sold a second time.

The wardens who actually did the punching weren't desk jockeys. They were more comfortable wielding a sharp knife than a hole punch. In 1948 the legal definition of a protected breeder was changed to a lobster with a "V-shaped notch" cut in her tail flipper, creating Maine's peculiar artifact, the V-notch. To control costs Maine also declared that the state would buy only female lobsters that had extruded their eggs *after* being caught. This limited the program to females that "egged out" while waiting in the holding pens of wholesale dealers before sale to retail outlets. The new rule benefited the dealers, but not fishermen.

Now the lobstermen stopped laughing. Lobstermen weren't desk jockeys either—a hole punch was about the last piece of equipment found aboard a lobster boat—but all lobstermen carried knives. Why should their tax dollars go to pay dealers and wardens to cut V-notches in lobsters before returning them to the sea? Any fisherman could catch an egger and cut her a notch. Perhaps from petty pride as much as altruism, in the 1950s the lobstermen of Maine began marking eggers with V-notches of their own volition.

When Warren Fernald's generation of lobstermen came of age and produced offspring of their own, they realized that every V-notch they cut was a deposit in the bank account of their children's future. By the time Warren's sons bought their first boats, the Maine lobsterman could legitimately claim to be a less murderous predator than his forebears. But the scientists in government, on a mission to protect the creatures of the sea from the rapacious hand of the fisherman, seemed not to know that.

~

"Ready about. Hard to lee!"

Bob Steneck was on vacation with his wife and parents,

steering a rented sailboat northeast along Maine's intricate coastline of islands and bays. Bob spun the wheel to starboard and cranked in the mainsheet while his father hauled in the jib. The boat leaned eagerly into the new tack. Bob glanced up at the mainsail and tightened the capstan half a turn until the sail quit luffing. He settled into the windward seat of the cockpit and wiped his damp beard on the sleeve of his shirt, glad to have gotten off the Darling Marine Center's leaky houseboat for a change. The August morning was unusually hot for Maine, and the bits of spray kicked up by the wind felt good. He peeked under the boom to check for lobster boats at work.

In midlife Bob's orange hair had begun its migration from the top of his head to his chin, but he was still athletic and his belt line was well under control. In high school his compact frame and muscular limbs had made Bob a tenacious wrestler and an agile soccer forward. While he hadn't been a stellar student, he'd been recruited to play college soccer and had made the best of the opportunity by excelling in physics, chemistry, and biology. He enjoyed competition and liked all his activities in life to be "goal oriented."

There was one exception. Bob and his wife, Joanne, an attorney for the state, were both hard workers, and sailing was their favorite indulgence. Bob's father had taught Bob to sail on Lake Hopatcong. With his parents now visiting Maine, Bob wanted to show them the spruce-covered islands and sparkling seas of his new home. The day before, they had sailed east across Penobscot Bay and anchored off the gentle slopes of Isle au Haut for the night. This morning they were continuing into the section of the Maine coast known as "Down East." The region had gotten its name from the prevailing breezes that were once so crucial to coastal commerce. Ships from Boston or western Maine sailed "down" the wind in order to travel eastward—really northeastward—up the coast.

Bob checked his chart. Looming ahead were the hills of Mount Desert Island. Bob was far from the first sailor to use these hills as a navigational aid. Visible from sixty miles out, the mountains of Mount Desert had been a landmark to generations of seamen before the Stenecks. Norsemen had sailed into the area perhaps a thousand years ago. European explorers, followed by cod fishermen, had sailed into the Gulf of Maine and used the Mount Desert hills as a navigational marker in the 1500s. Bob peered at his map again and saw that just south of Mount Desert, a cluster of small islands called the Cranberry Isles formed a protective anchorage. The harbor of Little Cranberry Island looked like a safe place to spend the night.

The first person to chart this group of small islands, in 1524, was the same man who discovered the bays off the island of Manhattan—the Italian explorer Giovanni da Verrazano. In Maine, Verrazano's landing party had been repelled by Native Americans wielding bows and arrows. In fury Verrazano had scrawled the name "Land of Bad People" across his map. When he left, the natives celebrated his eviction from their land—according to Verrazano's log—"by exhibiting bare buttocks and laughing." With luck Bob and his family would receive a warmer welcome.

Around noon their boat drifted into Little Cranberry's small harbor just as the wind died, making the day even hotter. After bowls of chowder at the restaurant on the wharf, the Stenecks strolled up the island's main street, past colorful flower gardens and lobster traps scattered in yards and driveways. Down a wooded road on the back of the island they passed a snug white house with blue trim, set back from the road at an angle. Heat radiated from the cracked pavement.

Emerging onto a beach facing the open ocean, they sat on the shore and rested, hoping to catch a breeze off the water. Bob snapped a photo of his family, who had towels draped over their heads against the sun. He laughed and stated that they looked like Bedouin fighters in a scene from *Lawrence of Arabia.* Except that they weren't sitting on sand. The beach was a mile-long arc of gray cobblestones, sloping down to the water and disappearing beneath the lapping waves.

Underneath those waves was one of the best lobster neighborhoods around. It was full of crustaceans, fishermen's traps, and lots of fights.

7

Battle Lines

Bruce Fernald glared at his landmarks, then looked back at the empty ocean.

"Come on, where are you?" he shouted. "You're supposed to be right here!"

Bruce hit the throttle and roared over the same splotch of sea once more, but he'd been back and forth so many times he was getting dizzy. Several of his buoys had simply disappeared.

"There's no *need*," he sighed, "of this unnecessary bullshit."

Recently a few lobstermen from the mainland had been tangling with the island fishermen over the boundaries of their trapping territories. Some were trying to evict others from the choice areas of bottom where the lobsters liked to congregate. Bruce had stayed out of it, but he had just learned that when there was a fight going on, no one was immune.

At the outset, such disagreements could be decorous, even polite. If a lobsterman anywhere on the Maine coast noticed that an intruder was setting traps over the traditional boundary, he followed a universal etiquette. First he snagged a few of the offending buoys with his gaff and retied them backward as a warning. If that didn't work, he might haul up the offending traps and throw them back with their doors open, or their bait bags removed. When the intruder failed to take the hint, the defender's last resort was to slice the buoy lines with a sharp knife. Lucrative lobstering territories were prized, and often fiercely guarded from one generation to the next.

Bruce was more inclined toward construction than destruction and soon discovered an outlet for his energy. By the early 1980s the lobstermen of Little Cranberry were eager for a reliable supply of the latest piece of lobstering technology—traps made of wire. The improved rectangular traps were constructed from plastic-coated metal mesh. Impervious to wood-eating worms, they required less maintenance than the old round-top traps made of lathes. Less time spent on maintenance meant a person could fish more gear and catch more lobsters. The Little Cranberry lobstermen had begun buying wire traps from a man on the mainland, but now the fellow had put his trap-building business up for sale.

"We ought to buy his equipment," Bruce suggested to his brother Dan.

"I wonder how much he wants for it," Dan said.

"Don't know," Bruce replied. "But I bet we could afford it if a bunch of us went in on it together. It'd be cheaper than buying traps or building them on our own."

Dan agreed, and with two other lobstermen the brothers pooled their cash. They hauled the rolls of metal mesh and the wire-bending machines out to the island and built a small barn outside Dan and Katy's house, where they installed their new lobster-trap factory. They called it the Cadillac Trap Company after Mount Desert Island's highest hill, and had a friend design a logo—a pair of lobsters driving a fin-tailed Cadillac convertible through the sea. On winter days when the weather was too rough to fish, the men would spend the day snipping, shaping, and snapping together sheets of wire. In the first year alone they constructed nearly a thousand traps. They patted each other on the back. A thousand traps could catch a lot of lobsters. The government scientists would have said too many, but the fishermen were ambitious, and now they had families to support.

Over the past decade, as the young lobstermen of Little Cranberry had matured, social life on the island had changed. For a time, the beach parties of the 1970s had grown wilder every summer. First marijuana and then cocaine had come to

Little Cranberry. Alcoholism was an ever-present threat, as it can often be in seasonal industries where months of intense work give way suddenly to periods of relative inactivity. Little Cranberry's remoteness made the long winters there especially difficult to endure—some mainlanders took to calling the island "a quaint drinking community with a lobstering problem."

A few of Little Cranberry's transplanted bachelors burned out and departed, and a few gutsy young women arrived to take their place, narrowing the gender gap. As the members of this generation paired off, settled down, and began to bear children, the parties gave way to domestic responsibilities and plans for the future. Not everyone made the transition successfully. Bouts of drinking and depression continued to plague pockets of the Little Cranberry community. Banding together to build lobster traps was a good way to kick the blues. The fishermen weren't just making traps. They were building faith in the future.

In 1983 Barb gave birth to a pair of identical twin boys. A year later Bruce bought a new boat, which he affectionately dubbed the *Double Trouble*. At nearly twice the size of the *Stormy Gale,* the *Double Trouble* gave Bruce the extra deck space he needed to handle more gear. He would have to make more money to pay off the investment, but in the long run he would come out ahead, assuming that the lobster catch stayed strong. There were risks, but he didn't dwell on them.

Bruce thought his expensive new boat was pretty impressive, until one afternoon he was overtaken by a green ship five times the size of the *Double Trouble*. Bruce's VHF radio blasted out a remarkable request from the ship's captain. The owner of the ship wanted the freshest lobster money could buy and insisted on making a purchase directly from Bruce's boat.

The crew lowered a bucket, and Bruce pulled alongside and stuffed it with the best of his day's catch. The bucket came back carrying far more cash than Bruce would have made selling to the co-op. When a gentleman of distinguished bearing leaned over the rail and threw a salute, Bruce knew that his efforts at trapping lobsters had just received the approval of one of the planet's richest men. For on the ship's flank was engraved in gold the name *Highlander,* which made the name of the man on board Malcolm Forbes.

~

Bob Steneck felt like a kid again. The world was new and exciting; fresh discoveries lay behind every rock. After years of research dedicated to the slow-motion consumption cycles of algae, snails, and sea urchins, the raucous energy of lobster life was intoxicating. Bob had also begun to realize that lobsters supported a vibrant industry involving thousands of hard-working families in hundreds of towns along the coast. And at the end of the day there was a huge bonus in the study of lobsters: you could eat them for dinner.

Instead of watching how quickly a sea urchin would devour a piece of kelp, Bob was now more interested in how quickly a lobster would devour a sea urchin. Back in the lab, Bob bent and shaped sheets of wire mesh into large cages with internal compartments and loaded them aboard his houseboat. His idea was to set the cages on the bottom, catch a few lobsters to put inside, and toss in urchins and other prey to see what the lobsters liked to munch on.

Out in the bay, Bob was muscling one of his wire-mesh cages over the side of his houseboat when he was startled by the roar of a diesel engine. A lobster boat had pulled alongside, white water boiling from under her stern as the captain brought her to a sudden halt a few yards away. The lobsterman asked Bob what he was doing.

"I'm doing some experiments with lobsters!" Bob shouted back, pleased to meet someone else as excited about the creatures as he was.

"Uh-huh," the man answered, not smiling. He stared at the wire cage Bob had in his hands. "You got a license for those things?"

"A license? Ah, I don't think I need one."

From the conversation that ensued, Bob would later

remember mostly the words "damn scientist" and "If you need to know anything about lobsters, just ask me." The fisherman reported Bob to the Department of Marine Resources, which ruled that Bob did, in fact, need a license to drop wire-mesh cages into some of the best lobstering territory in New England. State officials ordered that Bob's cages be hauled up off the bottom.

The episode wasn't the only brush Bob had with local authorities. When he heard that the scientists who worked for the state believed Maine's lobster population was in danger, Bob was dumbstruck. As an ecologist, trained to observe the abundance of organisms in their habitat, Bob couldn't help thinking that the scientists were wrong. From the countless hours he'd logged underwater, it seemed obvious that lobsters were wildly plentiful. Bob decided to pay a visit to the scientist in charge of lobsters at the Department of Marine Resources.

The visit did not go well. The scientist invited Bob into his office, and Bob described how many lobsters there were underwater, especially how many young lobsters he saw. It didn't look to Bob like a population in trouble. The scientist reached for a stack of papers and told Bob there weren't as many lobsters on the bottom as he thought. It wasn't a question of what he thought, Bob protested, it was a question of what he saw.

Years later, Bob would remember the scientist offering him a piece of advice: pick something else to study—lobsters wouldn't be around much longer.

~

Among the lobster boats moored in Little Cranberry Island's small harbor, Bruce Fernald's new boat, the *Double Trouble,* was giving Jack Merrill's *Bottom Dollar* a run for its money. Jack's green boat was still bigger, but the friendly competition between the island men had narrowed Jack's lead.

Jack wasn't worried. He was catching plenty of lobsters, and had also caught himself a wife named Erica. Jack and Erica had taken on parental duties as well, though at first it hadn't exactly been a child they were caring for.

One day a resident strolling along the island's rocky shore had stumbled onto an abandoned seal pup. The islanders knew they weren't supposed to approach it, but the fuzzy bundle was clearly near death. They carried it indoors and tried to feed it, to no avail. When Jack returned to the island after a day of lobstering, he tried a different approach. He called to the pup, vocalizing the sort of croak he'd heard mother seals making from the ledges around the island at low tide.

The pup stared at Jack. After a minute Jack decided there was nothing more he could do and turned to leave. He was nearly out the door when he heard an astonishing sound. The pup had responded to Jack with a call of its own.

The little seal couldn't be parted from Jack after that, and he secured an unusual federal permit to nurse the animal back to health, though not without some wangling. The agency in charge didn't believe a commercial fisherman could be entrusted with the care of a sea creature, so Erica's name was used on the permit instead.

Following instructions from the New England Aquarium, Jack took turns with Erica rising in the middle of the night to mix a warm cocktail of heavy cream, cottage cheese, puppy formula, and antibiotics in the kitchen blender. For good measure they threw in handfuls of chopped herring. If the concoction wasn't precisely the right temperature, the seal would spit it onto the linoleum.

Soon the sheen returned to the pup's coat. When the animal was strong enough Jack let it swim in tide pools at the water's edge. The seal threw fits if Jack left its side, so he took it lobstering with him in a garbage can fitted with running seawater. Within a month the seal was coming and going from the house on its own, flopping down the road to the beach when it wanted a swim and returning when it grew hungry. Soon its visits to the house became less frequent. The last Jack and Erica saw of it, the seal was bobbing a hundred feet off the beach, another seal at its side, looking at them before it dove and disappeared.

Lobsters might have been less lovable than baby seals, but

to Jack they weren't less deserving of human help. After studying marine biology in college, and after learning about lobster conservation from Warren Fernald and the other fishermen of his generation, Jack offered to take Bruce's place on the board of directors of the Maine Lobstermen's Association when Bruce's term expired in 1984. Jack believed that lobstermen were stewards, not exploiters, of the lobster population, and as an officer of the MLA he was prepared to fight the scientists in government to preserve his way of life. He believed that lobstermen could continue their intensive trapping, even increase it, and protect the resource at the same time.

What made Jack livid wasn't simply that the government scientists advocated raising the minimum legal size of lobsters. They also wanted to dispense with the very conservation practices that Jack and his fellow lobstermen believed were protecting the fishery—the V-notch and the maximum-size law.

The idea behind these practices was to build up a "brood stock" of large lobsters that would keep making eggs so the lobstermen could continue reaping their harvest of smaller lobsters. By cutting V-notches in females with eggs, the lobstermen were offering them a kind of reward card for getting pregnant. And by throwing back any lobster with a carapace over five inches, the lobstermen were populating a sort of sex resort for retirees, open to both male studs and experienced females. When a young female reached puberty, she could keep getting pregnant and earn several punches on her reward card, allowing her to retire to the sex resort for the rest of her days. Having secured membership in the lobstermen's brood stock, she might easily go on mating and making eggs for another fifty years. Indeed, for the male lobsters that made it to the sex resort, it was probably more like entering lobster heaven.

Ironically, the very question that the government scientists saw as a terrible conundrum—how could those lobsters, the eggers and oversize animals, possibly grow past adolescence in the first place?—was for Jack a wondrous mystery. The scientists' calculations suggested that a female lobster had to get really, really lucky to end up with a V-notch—she must reach puberty far ahead of schedule, have sex almost immediately, avoid traps until she extruded her eggs, then make sure she entered a trap during the few months she was carrying eggs. And yet Jack routinely witnessed shiny females, a few molts over the minimum measure and never notched before, coming up in his traps carrying eggs. Big males would come up with a carapace longer than five inches. Somehow, those lobsters had evaded capture despite the checkerboard of traps Jack and his friends set across the seafloor.

Jack couldn't say exactly why these conservation practices were effective, but he saw them working with his own eyes. He wanted desperately to show the government experts what he saw, but the scientists thought that information provided by fishermen would be biased, and they were unreceptive.

So instead, lobstermen decided to conduct their own research. The MLA began mailing postcards to its members every autumn, asking them to write down the numbers of eggers and V-notchers they caught. On Little Cranberry Island, Jack, the Fernalds, and a number of other lobstermen filled out their cards and mailed them back to the MLA office. It was easy for them. Katy Fernald had gotten them used to counting lobsters with her coffee cans.

By conducting their own survey, the lobstermen could at least collect a different kind of data from the scientists' data. All the same, Jack began to wonder if what Maine's lobstermen really needed was just a different kind of scientist.

~

Some seventy miles to the southwest of Little Cranberry Island, in the bay near the University of Maine's Darling Marine Center, word had spread among the local lobstermen that some scientist had been dumping cages that looked like traps into their lobstering territory. Several fishermen decided to investigate. A leader of the local gang of lobstermen and an MLA officer, Arnie Gamage, started to make it a habit to steer his lobster boat by Bob Steneck's houseboat to say hello. After they'd grown friendly, Bob made a remarkable request. He wondered if Arnie would take him out lobstering.

Bob's Ph.D. hadn't prepared him for the education he received aboard Arnie's boat, for Arnie put in the longest and hardest hours Bob had ever seen a man work. And after witnessing the ocean from a lobsterman's perspective, Bob discovered that a few things about his own experience began to make sense. Passing an inlet that was dotted with buoys, Arnie pointed.

"I sure would love to fish in that cove."

"So, why can't you?" Bob asked.

Arnie laughed. He was one of the most revered lobstermen on the water, but even for him the answer was simple. You didn't dump gear into someone else's territory unless you wanted to start a fight.

Arnie seemed to have put in a good word for Bob on the wharf, and soon he was on amiable terms with other fishermen too. One day Bob was invited to stop by the Maine Lobstermen's Association office after hours, where he found several of his new acquaintances and several six-packs of beer. Pretty soon Bob had drunk more than he should have, and pretty soon he was shooting the shit like one of the guys. In an epiphany inspired by Budweiser, Bob realized that if he planned to study lobsters on the Maine coast, this was the prerequisite.

One of the lobstermen told Bob that if he wanted to do research in their territory, all he had to do was ask. Because he was keen to observe ever better neighborhoods, it wasn't long before Bob had talked Arnie and his colleagues into removing their traps from a section of their best fishing ground so he could census the local population of lobsters. It was a feat unequaled in the history of lobster science, and it signaled a new era of collaborative research.

Bob's formal introduction to the MLA came later. In 1984, Bob was invited to give a talk on coastal carbon cycles at a scientific conference on the Gulf of Maine. From the podium he noticed a man sitting in the front row who looked different from the rest of the audience. With his jutting jaw and intense gaze, he had more the look of a fisherman than a scientist. Bob kept glancing at the man. The man stared straight back, absorbing every word.

After his talk Bob took a seat in the audience. In the talks on physical oceanography that followed, he could comprehend only half the technical jargon that his colleagues rattled off, but every time Bob stole a peek at the man in the front row, he was listening with rapt attention. From time to time the man would ask a question, his voice a determined coastal drawl. During a break Bob got to talking with the fellow, who introduced himself as Ed Blackmore, president of the MLA.

Ed Blackmore and his brother Frank had been raised by their grandfather, a lobsterman on Deer Isle, and had become lobstermen themselves before they'd had much choice in the matter. Ed was already trapping at the age of ten, and as he grew older the chip on his shoulder grew bigger. Ed felt the rest of the world disparaged lobstermen as second-class citizens. In 1954 he decided to change that by becoming a founding member of the MLA.

Bob's background couldn't have been more different, yet he quickly discovered that he and Ed shared an opinion: lobsters were abundant and probably weren't in trouble. Ed pointed out that by cutting V-notches in females with eggs, he and his fellow fishermen were protecting the supply of offspring. Bob described the many small lobsters he saw on the bottom, which he took as a sign that the population was healthy.

Months later, Ed invited Bob to make a formal presentation at the MLA's annual meeting of the board of directors. Bob wore a coat and tie, and brought along his first graduate student, a soft-spoken ecologist named Richard Wahle, who'd been helping Bob conduct his studies.

When Bob walked into the room full of tough-skinned fishermen in boots and jeans, his coat and tie didn't go over well. He launched into his presentation to a muted welcome, explaining that he'd discovered there were good neighborhoods for lobsters and bad neighborhoods. The good neighborhoods had more lobsters than the bad ones. Arms folded across their chests, the lobstermen shot each other glances across the table. Jesus Christ, this was science?

But then Bob showed some slides of lobsters underwater, and the fishermen liked that. At least Bob was actually looking at the bottom of the ocean, unlike most of the scientists they could think of. A few of Bob's comments about lobster behavior seemed off the mark, but on the whole the men didn't think this ebullient biologist could do them harm.

Afterward a young bearded man with a ready smile shook Bob's hand and thanked him for coming.

"I didn't agree with everything you said," Jack Merrill commented, "but it's the first time I've heard a lobster scientist say anything that made any sense."

~

When they'd moved to Maine, Bob and Joanne Steneck had fallen in love with a quirky, sprawling house on a wooded hillside carpeted with fern, half an hour inland from Bob's lab at the Darling Marine Center. The heart of the home was a big living room like an atrium, with floor-to-ceiling windows two stories high, looking out on a grove of oak and beech trees. While Bob often enjoyed the relaxed fit of the spacious living room, sometimes he preferred the restricted fit of the kitchen, where he liked to curl up and read the paper in a cozy L-shaped breakfast nook. He had backed himself into the bench one Sunday morning in 1986, coffee in hand, when he saw the editorial.

The *Portland Press Herald,* the state's largest newspaper, had accused Maine's lobstermen of resisting the government's advice out of short-term greed. The scientific community was unanimous, the editors wrote, in its determination that the minimum legal size of lobster had to be raised. That would allow the lobster population to expand to meet the pressures of increased trapping.

"Whoa!" Bob snorted, almost spilling his coffee. "What?"

First, most of the lobstermen Bob had met didn't fit that description. Second, Bob was a member of the scientific community, and he didn't think Maine's lobsters were necessarily in trouble.

From his studies of lobster neighborhoods, Bob knew that the newspaper's assertion wasn't a foregone conclusion. Given how territorial lobsters were, Bob suspected that the number of nooks and crannies on the bottom might determine the number of animals that could live in a given area. Most lobsters, especially the young, needed shelters with the right fit. If they couldn't secure one, or if the neighborhood was too crowded, they'd search somewhere else. Bob had also seen that the geology of the coastal seafloor where he worked limited the number of good neighborhoods. Bob would have to saw up an awful lot of PVC pipe before sediment flats or open bedrock became desirable real estate for lobsters. And, come to think of it, the notion that just adding eggs would result in more lobsters was like arguing that a farmer could fend off a drought by dumping more seed on his field.

In Maine's recorded history, the annual catch of lobsters had never much exceeded twenty million pounds, no matter how many lobstermen were on the water and no matter how many traps they fished. In 1950 there were five thousand registered lobstermen in Maine and five hundred thousand traps, and the catch was about twenty million pounds. By 1974 there were eleven thousand lobstermen and two million traps, but the catch had dropped to sixteen million pounds. As more traps were added, however, the catch rebounded to about twenty million pounds, and there it stayed.

Canadian biologists had noticed a similarly strange phenomenon near Prince Edward Island in the 1950s and 1960s. In an attempt to predict future catches, for nearly two decades they had towed a small net through the waters of Northumberland Strait every year and recorded the number of lobster larvae they caught. Some years the strait was teeming with larvae; other years it contained only a modest amount. But seven years later—approximately how long it took lobsters to mature to harvestable size—greater amounts of larvae had never consistently resulted in a larger haul of lobsters. As with Maine's unvarying catch, the Canadian data suggested that a fixed amount of shelter on the seafloor capped the number of new lobsters, year after year. How this happened exactly wasn't clear. But like a game of musical chairs, when a lobster couldn't find a place to sit, apparently it lost.

Bob made himself another cup of coffee. Newspaper in hand, he climbed a ladder to his office, a kind of glassed-in tree house on the edge of the building, where he did his best thinking. Perched at his desk, he read the editorial again. He snatched up a pen to write a rebuttal.

8

The War of the Eggs

Bob Steneck didn't know it, but this wasn't the first time scientists and fishermen in Maine had clashed over the minimum size of lobsters. The first fight had occurred nearly a hundred years earlier, and the combatant on the side of science had been another lobster biologist trained at Johns Hopkins University. Where Bob Steneck was inclined to side with the lobstermen, Francis Herrick, the scientist who had solved the mystery of the male lobster's dual penis, had been the lobstermen's harshest critic.

For most of the nineteenth century, lobsters couldn't be transported alive because there was no refrigeration. The animals died before they reached market, and the meat of a dead lobster quickly developed toxins. For the fishermen of the Cranberry Isles and other remote communities in eastern Maine, there was no point in catching lobsters until the second half of the 1800s, when canning factories were built Down East, including several on Mount Desert Island. Lobster meat could be boiled and shipped in sealed tins. It didn't matter what size a lobster was because the cannery workers picked the meat from the shell. At first the fishermen caught mostly large lobsters, but soon the big animals became scarce and the fishermen increasingly relied on young lobsters to earn their pay.

In the 1870s construction crews laid railroad tracks into Maine, linking the western half of the Maine coast to the metropolises of Boston and New York. Lobstermen who lived near the railroad could pack lobsters in ice and ship them live to many parts of the country—one story credits the newspaper tycoon William Randolph Hearst with the first order for a dinner party in Colorado. But to satisfy diners like Hearst, the fishermen needed lobsters large enough to fill a dinner plate. The tiny lobsters caught for canning wouldn't do.

Faced with a dwindling supply of large animals, the lobstermen who worked along the western part of the Maine coast declared war against their fellow fishermen Down East. The western lobstermen touted a minimum-size rule to allow lobsters to mature to a more lucrative length. The eastern lobstermen lived in a distant part of the state unreachable by train, and they opposed the rule. The fight between the live-lobster fishermen and the cannery fishermen dragged on for years, each side trying to put the other out of business.

The legislature enacted a string of dubious half measures until 1895, when Francis Herrick published his report, *The American Lobster: Its Habits and Development*. Even the male lobster's dual penis wouldn't be sufficient to repopulate the stock, Herrick feared, if the fishermen Down East continued to strip lobsters from the sea before they had a chance to reach puberty.

"The lobster may be rightly called the King of the Crustacea," Herrick wrote, "in consideration of both its size and strength, its abundance and economic value." But Herrick went on to state that the industry was in trouble, noting "the gradual diminution in the size of the lobsters caught and an undue increase in the number of traps and fishermen." Herrick predicted that fishermen would drive the lobster to commercial extinction. "Civilized man," he concluded, "is sweeping off the face of the earth one after another some of its most interesting and valuable animals, by a lack of foresight and selfish zeal unworthy of the savage." Perhaps inspired by Herrick's rhetoric, the legislature finally voted to give a minimum-size rule the force of law.

When he was a graduate student at Johns Hopkins University, Bob Steneck's favorite refuge had been a wood-paneled reading room lined with old dissertations. Buried in his own research, Bob had been oblivious of the original copy of Francis Herrick's report, which sat just feet away on the shelf. Had Bob pulled it down and dusted it off, he might have read Herrick's portentous words and remembered them. He might have gained an inkling of the stakes in this fight, and he might have thought twice about getting involved.

Electro-ejaculation wasn't pleasant. The big male fought valiantly while Susan Waddy strapped him upside down to the operating table. Once she had secured the wide rubber flaps over his tail, thorax, and claws, the lobster was immobilized, and he stopped struggling. Next to the table she had readied a transformer, voltage meter, and electrodes. A large female lobster was already strapped upside down beside the male. Susan set the regulator to deliver a burst of 10-milliamp alternating current.

Susan Waddy and her fellow lobster scientists at the St. Andrews Biological Station in New Brunswick—the station was part of the Canadian Department of Fisheries and Oceans—had learned from trial and error that alternating current worked better than direct current. Voltage mattered too. "Use of an ammeter is recommended," they noted in their report, "to prevent problems that can result from stimulation with excessive or inadequate current."

Susan squeezed open the toothed claw of the negatively charged electrode clip and secured it at the base of the male lobster's pair of hard, penislike swimmerets. The positive electrode was a blunt probe two millimeters in diameter, which she placed on the lobster's belly next to the opening of the sperm duct. With her other hand she pressed the switch on the power supply and delivered the electro-ejaculatory jolt. There was a spasm and a perfect sperm packet popped out. Susan exchanged the electrode for a pair of forceps and lifted the intact spermatophore from the male. With a second pair of forceps she forced open the doors to the female's seminal receptacle and plopped the sperm packet inside. The artificial insemination was complete.

The inseminated female was one of twenty female lobsters that had been living in tanks at the St. Andrews station for more than a decade. This female hadn't retained the sperm she'd originally received during a natural copulation, so Susan had re-inseminated her.

In general, though, Susan was amazed at how long the old females could retain sperm. The St. Andrews animals were the only female American lobsters in captivity whose reproductive abilities had been under scrutiny for so many years. By now these were big lobsters—some over two feet long. What Susan and her colleagues had discovered was that as the females aged, they became more adept at mating and making eggs. This was important, because many scientists assumed female lobsters grew less fecund with age.

A female lobster's eggs develop for nine or ten months inside her ovaries. Then she finds a secluded spot, lies on her back, folds her tail to create a kind of basket, and squirts the eggs out through a pair of ducts. At the same time, she unzips her seminal receptacle and fertilizes the eggs with the sperm she has stored since mating. Then she attaches the eggs to the underside of her tail, using a glue she produces from cement glands on her swimmerets. She carries the eggs around for another ten months or so, allowing them to develop before they hatch, at which point she finally releases them.

As early as the 1890s, Francis Herrick had discovered that older female lobsters produced more eggs than younger ones. Over several consecutive summers at his lab in Woods Hole, Herrick counted the tiny eggs glued to the tails of four thousand female lobsters. He discovered that a lobster's capacity for egg production increased exponentially with her size. A small female that was eight inches from nose to tail could extrude about five thousand eggs. A lobster twice that length extruded ten times as many eggs, around fifty thousand. Herrick even found one female that was carrying more than ninety-seven thousand eggs.

Subsequent scientists, however, noted that large females molted much less often than small females. Since mating occurred during molting, it followed that large females would spawn less frequently than smaller females, canceling out much of the advantage of the extra eggs.

But Susan Waddy and her colleagues in Canada had turned this thinking on its head. The St. Andrews lobsters revealed that veteran females develop tricks in the battle of the sexes that younger females can't match. As a female grows older, her seminal receptacle matures from a simple pouch into a sperm bank that can accommodate more spermatophores and preserve them for several years. After copulating once, an older female can produce and fertilize two entire batches of eggs without bothering to molt or mate a second time. Susan's veteran ladies needed a man around only once every four or five years, but they still produced eggs more often and in vastly greater quantities than their smaller counterparts. A rough calculation suggested that over a period of five years, one five-pound female could produce as many eggs as twenty-seven one-pound females.

On Little Cranberry Island, Jack Merrill was in need of firepower for his battle against government scientists. The publication of Susan Waddy's findings in 1986 hit with the force of field artillery. Jack wasn't about to suggest that fishermen set up electro-ejaculation operating tables aboard their boats. Nor did Susan Waddy's one-to-twenty-seven ratio exactly apply in the reality of Maine waters, where cold temperatures prevented most females from reaching sexual maturity at a weight of one pound. But the gist of the discovery was crucial. In their confrontations with the government, Jack and his colleagues in the Maine Lobstermen's Association had been arguing for the benefits to egg production brought by big lobsters—the animals protected by V-notching and the oversize law—and so far the arguments had fallen on deaf ears. The government scientists insisted on boosting egg supply by protecting smaller lobsters with an increase in the minimum size. To Jack that didn't make sense, especially if one woman could make as many eggs as a whole gang of girls.

Jack Merrill had driven to a meeting of lobstermen, government scientists, and state officials, and now he was waiting for his turn to speak. He slipped a stack of papers from his bag and thumbed through them for the relevant sections. He was ready to launch a surprise attack.

Jack had unearthed more than just Susan Waddy's paper on female fecundity. In 1985 the Maine legislature, bewildered by the battle between government scientists and lobstermen, had hired a team of outside consultants to study the debate over lobster management in Maine. The lead author, Louis Botsford, was a specialist in population dynamics from California. Botsford had submitted the report to the Department of Marine Resources in the spring of 1986. The sections of the report the department had made public supported the government's argument that the minimum size of lobsters needed to be raised to increase egg production. That made Jack suspicious. He asked a friend on Little Cranberry to see if she could obtain a copy of Botsford's entire report when she passed through the state capital.

In Augusta a few days later, Jack's friend and her teenage son, a budding Little Cranberry lobsterman with his own skiff and traps, were turned away when they asked to see the report. They tried a different office and were told that only one copy of the report existed, and it couldn't be copied. Losing patience, they mentioned that this young lobsterman happened to be friends with the son of the majority leader in the state house of representatives. That produced ten copies, one of which they took and passed on to Jack Merrill. For the young fisherman and his mother it was a sobering civics lesson. For Jack it was a revelation about the political uses of science.

When Jack stood up to speak, the state's chief lobster biologist had just finished citing the Botsford report to bolster the government's argument for raising the minimum size. Now Jack cleared his throat.

"I'd like to read you a few sentences from the Botsford report that you haven't heard yet, and they tell a somewhat different story."

Jack began to quote from unpublished portions of the report:

"'The V-notching program holds substantial promise as a means of protecting the brood stock. If we assume for the sake of comparison that one out of every four un-notched egged females that is caught gets V-notched every year, then total egg production will be more than doubled for only a slight decline in catch.'"

The room erupted. The lobstermen leaped to their feet and gave Jack a standing ovation.

Jack felt as if he were trying to pull lobster science out of a deep, dark hole. How had things backtracked so far? Many decades earlier, it was the government that had used science to insist that lobstermen protect large lobsters instead of small ones, not lobstermen trying to convince the government of the same thing, as they were now.

Back in 1895 even the minimum-size law hadn't been sufficient to save the faltering fishery. Francis Herrick returned to Maine in the early 1900s and found that the region Down East remained a lawless backwater. Fishermen were chopping undersized lobsters up for bait to attract larger lobsters, or taking the little lobsters home in sacks to boil and sell as picked meat. Boats from Connecticut sailed among the islands incognito, purchasing shorts and transporting them south, where small lobsters were still legal. A thriving black market in Massachusetts also encouraged the smuggling of undersized lobsters down the coast by truck, train, and boat. Seeing that little had changed since his pessimistic assessment in the 1890s, Herrick believed the American lobster was still teetering on the brink of commercial extinction. He told officials in Maine that the minimum-size rule was ineffective. Drastic measures were necessary.

Aware that a female lobster produced exponentially more eggs the larger she grew, Herrick proposed a novel compromise. Why not actually lower Maine's minimum legal size to allow lobstermen to sell some of the small lobsters they were keeping anyway, and compensate by adding a maximum size to protect big lobsters? Given Maine's chilly waters, the large lobsters were the ones with the potential to rebuild the supply of eggs, not the small ones.

Maine's commissioner of fisheries endorsed Herrick's plan. To generate goodwill with lobstermen, the commissioner whipped up a pro-lobster frenzy, urging the people of Maine to eat lobster twice a week as a public service. Then he gave fishermen the bad news: "You are murdering your own industry."

Lobstermen ignored the warning even as catches continued to plummet. On Little Cranberry Island, when Warren Fernald's father, Malcolm, was seventeen and lobstering on his own, catches were miserable—in 1919 the state's annual yield dropped below six million pounds. Nevertheless, Malcolm and the other Little Cranberry fishermen set and hauled their poverty crates with stoic determination.

When Malcolm Fernald turned twenty-one the island was beset by a terrible omen. It was the coldest winter anyone could remember, and in February the ocean froze all the way from Little Cranberry to Mount Desert. The fishermen stumbled wide-eyed down the harbor, stepped onto what used to be water, and wandered around their locked-in boats in disbelief. Some of them walked all the way to the mainland. Robbed of what set the island apart—the ocean—Little Cranberry lost its hold on its people.

The omen foretold the Great Depression. The price lobstermen received for their meager catch plummeted, and nearly a third of the fishermen in the lobster trade left to find other employment. Little Cranberry lost nearly an entire generation of lobstermen as the island's children scattered to distant corners of the American mainland to search for work. Recognizing the disaster for what it was, in 1933 a slim majority of lobstermen acquiesced to Francis Herrick's recommendation, and the legislature finally enacted Herrick's maximum-size limit into law, affording new protection to large lobsters.

Lobstermen, Jack Merrill felt, had learned their lesson. Having suffered defeat, fishermen now endorsed protections for the lobster population. They had learned to V-notch, throw back oversize males and females, and give little lobsters a free lunch until they reached the legal size for harvest. But now where was Jack? Standing in front of government scientists who wanted to roll back the hard-won protections that fishermen had finally embraced, reminding the experts of the science that they now refused to accept.

~

While Jack Merrill collected scientific evidence, Ed Blackmore, the president of the Maine Lobstermen's Association, talked with fishermen and collected opinions.

For many lobstermen, raising the minimum carapace length to three and a half inches, as scientists were recommending, sounded like suicide. Personally, Ed was willing to consider a significant increase, but he worried about losing the market for small chicken lobsters to Canada. There was a flip side to a size increase, however. A lobster that was only slightly larger would still be marketable, and selling the animals when they were a tad heftier would boost the average lobsterman's income. About half the fishermen Ed canvassed opposed any change, and about half thought a small increase might be lucrative. Government economists thought the whole debate was crazy. Their research showed that people who bought lobsters were happy to pay more for the extra meat in bigger animals. Over the short term, lobstermen might lose money while waiting for the first generation of lobsters to grow larger, but after that they would profit in proportion to the increased size of the animals.

Like Francis Herrick before him, Ed Blackmore proposed a novel compromise. If the other New England states and Canada agreed to a corresponding hike in their minimum size—so no one gained the advantage of being able to sell cheaper chicken lobsters—then Maine would go along with a partial increase. But in exchange, Ed made an audacious demand. He asked that protection for V-notched lobsters remain not just in effect, but that it be expanded. He demanded that Maine's sacred females be allowed to roam free of molestation throughout all of New England's waters and out to the limit of United States federal authority, two hundred miles from shore.

The proposal seemed preposterous, yet the government could no longer dismiss it as the product of ignorance. The research that Jack Merrill had helped unearth lent credibility to the reproductive power of V-notched and oversize lobsters, and by now the MLA's postcard survey had amassed three years of data. For a two-day period every autumn, more than a hundred MLA members around the state had recorded the lobsters that came up in their traps. Statisticians at the University of Maine had analyzed the data and concluded that during the survey periods, nearly a third of all female lobsters that fishermen hauled up carried V-notches. That was a lot of egg-producing power.

Also bolstering the MLA's case were the arguments of a renegade scientist named Bob Steneck—the lobstermen's new ally. The *Portland Press Herald* had published Bob's letter to the editor next to a cartoon of a flustered lobster in a steaming pot. Calling himself a lobster ecologist, Bob had written that the lobster population might not, in fact, be in trouble at all. "We should be honest with ourselves and with the lobstermen," Bob's rebuttal had read, "and suggest that increasing the minimum size might be a prudent thing to do, but it should not be asserted as a scientifically deduced conclusion." If the need for a size increase was questionable, as Bob claimed it was, then few lobstermen would be willing to make the sacrifices the government's plan could entail.

Tough lobbying by federal officials convinced the Canadian government to commit to the deal, and despite misgivings on all sides, Ed's compromise plan was accepted in 1987. When the details were hammered out, the minimum legal carapace length for lobsters landed in Maine would go up by only an eighth of an inch—less than half of what government scientists had wanted.

It seemed a tiny change. But for lobstermen, whose annual

income depended on minuscule differences in the shell size of millions of newly molted lobsters every season, that eighth of an inch was likely to pack a wallop. Suddenly whole swaths of the lobster population would be off-limits, and the lobstermen would have to wait for the animals to molt again before they could be sold.

To ease the economic strain on fishing communities, the increase was set to phase in over four annual increments beginning in 1988, with a year off in the middle. And so it was that protection for V-notched lobsters went into effect throughout state and federal waters off New England, and Maine lobstermen agreed to trade in their old brass gauges for longer ones.

Government scientists believed few lobsters would ever earn V-notches, while lobstermen believed they would see many that did. Either way, it was a victory for the lobster industry, and perhaps the only instance in the history of commercial fishing in which fishermen had agreed to a new government restriction only if they could keep an old one too. Maybe that was what made it a deal that could never last. That, or the fact that within a year, Bob Steneck got a terrible idea.

9

Claw Lock

Two lobsters in a confined space will approach each other fearfully at first, and may leap backward when they come into initial contact. But their caution doesn't last.

The lobsters circle and slash at each other with their antennae. This fencing match emboldens them and they slice their claws through the water, from a widespread position to a crossed position and back out again. Next they press their claws together and shove each other back and forth, like a pair of prizefighters caught in a belligerent embrace. If one of the lobsters is larger than the other, and the smaller one can find no escape and no place to hide, the contest often ends with an abject display of groveling by the weaker animal. But if the two lobsters are evenly matched, then the shoving settles nothing. The fight escalates to a new level of tension and danger: claw lock.

Lobsters begin life ambidextrous, their two claws identical in shape and size. During their first year or two they start to favor either the right or left claw for crushing and the other for seizing and cutting, thus becoming either right- or left-"handed." The lobster's body also develops two basic types of muscle—fast fibers, which produce rapid contractions but tire quickly, and slow fibers, which produce gradual contractions of greater strength and longer duration. The lobster's tail, for example, contains fast fibers, useful during the escape reflex of swimming in backward bursts. The walking legs, which are used to jog at average speeds for long periods, develop slow fibers. Similarly, the seizer claw fills with fast muscle and remains streamlined while the crusher claw fills with slow muscle and becomes bulky, like a bodybuilder's bicep.

To test each other's brawn and slow-muscle stamina, the dueling lobsters desist from their shoving match and grip each other's crusher claws. If the lobsters are both right-handed, they reach across to grip each other as though shaking hands. If one of them is right-handed and the other left-handed, instead of reaching across they hold hands on the same side. And then they squeeze.

By entering claw lock the lobsters have traded their showy bout of shoving for a calmer but more consequential contest. It is a battle of endurance and a game of chicken—each of the lobster's shells straining under the pressure but neither combatant willing to ease up. After fifteen to twenty seconds, one of the lobsters will usually attempt to retreat before its shell shatters, and the winner releases its grip.

The lobster's repertoire of escalation helps avoid unnecessary injury. Most fights in the wild are settled without actual violence and end when one of the lobsters runs away. However, if both lobsters are similar in size and equally aggressive, even claw lock may be insufficient to settle the contest. If the fight escalates further, the claws become terrible weapons of destruction. One or both of the animals is bound to lose an antenna, leg, claw, or eye.

If one lobster gets a grip on an appendage of the other, the trapped lobster may jettison its compromised claw or leg by means of a special muscle at the base of the appendage designed to slice off that lobster's own limb. This capability, called autotomy, serves as an escape mechanism. But autotomy, like amputation, is also a kind of field-hospital first aid. Lobsters have open circulatory systems, meaning that their blood flows through body cavities rather than inside veins. A leak in a lobster's shell, such as that caused by a crushed or punctured leg, can cause the animal to bleed to death unless it cuts off the limb and seals the joint. A blood leak is a liability for another reason. Generally lobsters are not cannibalistic, but the scent of an injured lobster's blood can inspire them to kill and consume their unlucky comrade.

With time, lobsters are able to regenerate most appendages, although the energy required to do so slows their overall growth. An eye, unfortunately, will never grow back. But other appendages may appear grotesquely in the eye's place—an unwanted foot, for instance.

If one of the lobsters capitulates before being destroyed, it may receive mercy. A generous victor will pursue the loser to ensure that it assumes a submissive posture by folding back its antennae, lowering its claws, curling its tail, and backing into a corner. The winner will strut away, satisfied that it has achieved dominance. A spiteful victor, on the other hand, might chase the loser down and hack it to death.

Given the hazards of lobster fighting, Bob Steneck wasn't surprised that in his experiments with neighborhoods of plastic pipes, smaller lobsters moved out when large lobsters moved in. Lobsters were better off abandoning their homes if the alternative was to stay and fight. However, when he'd moved his pipes closer together, Bob had seen that even big lobsters would leave the neighborhood rather than spend all their time fighting overwhelming numbers of small lobsters.

As an ecologist, Bob had been trained not only to measure the abundance of organisms but also to map their patterns of distribution, and lately he had been nursing a horrible thought. What if his neighborhoods of pipes were a microcosm of the entire Maine coast?

The minimum legal size of lobster had already begun to go up, following the deal the lobstermen had struck with government scientists. Never mind that like most lobstermen, Bob hadn't thought the size increase was necessary in the first place. Now he was worried about the effect the size increase might have on the distribution of lobsters underwater. As the increments of the increase went into effect, the waters off Maine would be teeming with millions of slightly bigger lobsters—lobsters that in the past would have ended up on dinner plates. These lobsters would now take over the rocky bottom near shore, while the lobstermen would try to catch animals one molt size higher. But if Bob's pipe experiments were any indication, the lobsters one molt size higher might abandon their old neighborhoods rather than spend all their time fighting the influx of new residents.

Fishermen and scientists alike had observed that young lobsters tended to stay close to shore, while mature lobsters spent more time in deeper water. If the minimum-size increase caused overcrowding along the coast, Bob worried, the natural segregation of lobsters by age could be amplified. The larger lobsters the fishermen would now be pursuing might move further away from shore.

Most of Maine's lobster catch was trapped within a few miles of land. Compared with the big trawlers or swordfish boats that sailed out to sea with four-man crews for a week or two at a time, lobstermen were small operators. Most lobstermen owned their own boats and returned home every evening. They knew each other and fished inside fixed territories, which contributed to their culture of conservation. Bob worried that now, in the name of conservation, the size increases could push the new legal lobsters far enough from shore that Maine's lobstermen might have trouble catching them at all.

In their battle with the government scientists, Maine's lobstermen had won a vast territorial expansion for the protection of their big brood-stock lobsters. Federal authorities had turned some sixty thousand square miles of ocean over to the lobstermen of Maine as a sanctuary for wild lobster sex. One might think that having attained so much, the lobstermen wouldn't begrudge their adversaries a distance smaller than the diameter of one drop of water—a sixteenth of an inch. Yet they did.

In 1987 officers of the Maine Marine Patrol had ordered boxes of shiny new brass gauges, certified them for accuracy, and distributed them to supply shops along the coast. On the first day of 1988, lobstermen had traded up. The new rulers were bigger by one-thirty-second of an inch—about the width of a fingernail file. The fact that the difference was nearly imperceptible to the human eye didn't stop fishermen from complaining. By the time the gauge was lengthened by another thirty-second of an inch on the first day of 1989, the chorus of complaints was becoming a cacophony of discontent.

The scientists in government were baffled by the noise. The fishermen's initial fears—losing a portion of their catch and losing market share to Canada—had so far proved unfounded. In the year following the first increase, Maine's lobster catch hadn't fallen; on the contrary, it had overflowed by two million extra pounds. And even after the second increase, the portion of chicken lobsters had dipped from 66 percent to 63 percent of the catch—hardly a decline that would surrender the market to Maine's neighbors to the north. Two similar increments remained to be implemented in 1991 and 1992, so there was only a sixteenth of an inch left to go.

All the same, many lobstermen decided they'd had enough. Given Maine's cold water, some questioned whether a significant payoff in egg production could be achieved by the additional sixteenth of an inch. Others noted that Canada had yet to implement its part of the bargain by raising the size of its lobsters to match the new U.S. minimum. Still others feared that even with Canadian cooperation, consumers might balk at buying bigger lobsters in the years to come. With these concerns as a backdrop, Maine's lobstermen set the stage to renege on their agreement with the government. Complacency and stubbornness played a role too. The agreement had hardly been given a fair chance.

Leaving his coat and tie in the closet, Bob Steneck had continued to visit the MLA's board meetings to give the fishermen updates on his research. During the winter of 1989, after the second size increase, Bob reminded the members of the MLA that the number of good neighborhoods probably limited the number of lobsters, regardless of how many eggs were produced, so the size increase might be of questionable benefit. But Bob voiced a fresh concern as well. Based on another aspect of his research, he warned that too much of a size increase might spark a kind of gang war along the coast. Underwater, millions of slightly larger lobsters might be trying to push, shove, and box their way through crowds of smaller lobsters in search of shelter. Overwhelmed, the new legal-size lobsters might be giving up and moving into deeper water.

The lobstermen in Bob's audience had spent years, if not decades, carving out territory in which to set their traps. Fragile social compacts regulated which cove, ledge, or underwater gully was shared by whom. An additional sixteenth of an inch wasn't much, but if the new legal lobsters moved even a few miles from traditional fishing grounds, the scramble to realign territorial boundaries would be a nightmare. And if the new legal lobsters did migrate into deeper water, the costs to lobstermen would include longer days of hauling and cruising, steeper fuel bills, lengthened ropes, and the danger of heavier seas.

It didn't take long for the fishermen to see that the two prongs of Bob's research formed a powerful pincer. It was a pincer that might be able to put a stranglehold on that last sixteenth of an inch, and they were seized with hope. Within weeks Ed Blackmore had submitted a petition to delay the remaining two increases in the minimum size. Bob typed up a paper summarizing his research, which Ed added to the petition.

Bob returned to his busy schedule of teaching at the University of Maine, but a couple of months later he received an invitation to discuss his work in person before the committee of government scientists that oversaw lobster management. Pleased, Bob rose early one May morning, picked up a couple of his graduate students, and drove down to the committee's headquarters in Massachusetts. Bob enjoyed giving talks, and today he was hoping to have some fun. He'd brought along videos of lobsters fighting over his pipes—always a crowd-pleaser—and handouts with easy-to-read data. By the end of the day, the committee might even thank Bob for saving the fishery.

When Bob walked into the conference room he encountered

a forbidding court of experts. Around the table sat some two dozen lobster scientists, arms folded across their chests, expressions grim. Bob cracked a couple of jokes to break the ice. There was silence. The situation did not improve as he played the home videos from his houseboat. He explained that like the lobsters in the videos, the larger animals created by the size increase might be moving out of their old neighborhoods as lots of smaller lobsters moved in.

Bob had failed to realize that while he'd been playing with plastic pipes on the bottom of the ocean, the cogwheels of an unwieldy government apparatus had been gearing up to produce a law. During the 1980s, years of labor by dozens of talented biologists, mathematicians, economists, and managers had finally resulted in a rule raising the minimum legal size of lobster. For those promoting it, the legislation was especially worthy. Its goal was to ensure the future health of two types of sea creatures, lobsters and lobstermen, neither of which seemed to know what was good for it. The government had studied the position of the lobstermen and, after taxing negotiations and painful concessions, had reached an agreement. Bob had studied the position of the lobsters, and now the agreement was threatening to come undone.

The committee of experts listened carefully to Bob's presentation, but decided there was a big problem with the state of his science. It hadn't been peer-reviewed. Usually scientific findings must be vetted by colleagues in the field and then published in a respected journal before being applied to policy.

The committee's solution to this problem was to perform its own peer review. After hours of grueling interrogation and counterargument, instead of being thanked for saving the fishery, Bob was told that the committee would issue an official finding forthwith. And then there was the three-hour drive home.

~

It was early August 1989, and several hundred lobstermen had shown up for the fisheries management meeting in

Massachusetts, an unusual number considering that the peak trapping season was under way. Government officials were scheduled to announce a verdict on Bob Steneck's research, and to vote on the MLA's petition to delay future size increases.

Jack Merrill had driven down from Maine, expecting the vote to take only an hour or two. He looked at his watch. He'd now been waiting in the stuffy conference room for nearly ten hours. The officials in charge had introduced one delay after another, and the number of lobstermen in the room had dwindled to sixty or seventy. The fishermen who remained were getting rowdy. Finally a government scientist cleared his throat, and Jack listened for the verdict.

"Dr. Steneck's work," the scientist intoned, "does not provide sufficient scientific evidence to advise terminating the gauge increases."

When the vote by management officials was held a few minutes later, the MLA's petition was soundly defeated. Shell-shocked, Jack and his fellow fishermen filed out of the room and dispersed to their villages along the coast. They had, it appeared, lost their fight.

Then something strange happened. Up north, Canadian fishermen got wind of the controversy. They didn't like what they smelled and slammed the door on the whole deal. Suddenly, the Canadian government's agreement to match U.S. size increases was dead in the water.

Sensing an international crisis, the governor of Maine, John McKernan, and U.S. Senate majority leader George Mitchell, a Democrat from Maine, contacted the White House. President George H. W. Bush met with Canadian prime minister Brian Mulroney at the Bush family's summer home in Kennebunkport, but last-minute negotiations failed to save the agreement. When the president dined at Mabel's Lobster Claw that week he ordered prime rib. Lobster, the president was heard to complain, kept him awake at night.

He wasn't the only one. Ed Blackmore was so agitated that he gathered his counterparts from the Massachusetts and Rhode Island lobstermen's associations and called for state legislatures to circumvent the government scientists and declare a moratorium on size increases. Canadian cooperation had been a clear requirement. Without it, American lobstermen felt they'd been hoodwinked. They rallied and flew flags picturing a brass ruler with a red line through it.

"You know us Mainers," Ed told a reporter, "we grab onto something, we're not going to let go."

His opponents squeezed back. In the press, government experts questioned Bob Steneck's credibility as a scientist, and the *Portland Press Herald* ran another editorial, this time criticizing Bob and his irresponsible theories. Bob, gripping a lobster by the claws, was photographed by the Associated Press in a conspiratorial huddle with Ed. The picture ran in the *Washington Post,* above a quotation from a government biologist implying that thanks to Bob Steneck, the American lobster might now go extinct.

But the government's counterattack came too late. State legislatures sided with the industry and voted to block future increases in the minimum size. The lobstermen had won.

It wasn't clear to Bob that the turn of events represented any sort of victory for him. Like a big lobster surrounded by an overwhelming number of small lobsters, Bob could now spend all day fighting, or he could go somewhere else. Bob was up for a sabbatical the following year, and he decided to get as far away from Maine as he could.

He struck out for Chile, where he taught evolution at Catholic University in Santiago and studied a rare freshwater crustacean called the squat lobster. Joanne flew down and they took a couple of weeks off to trek in Torres del Paine, with South American camels called guanacos carrying their packs. Bob then traveled to South Africa, where he indulged in some research on seaweed-eating limpets at the University of Cape Town, and helped out with a classification study on the diminutive cape lobster. It was on a visit to Krueger National Park, staring a lion in the face, that Bob realized he finally felt like a kid again. When Bob's professor in college had beckoned him into the underbrush in search of wood ducks, it had defined science for Bob as the joy of discovery. He'd almost forgotten what that felt like.

And yet strolling South Africa's spectacular sand beaches, Bob couldn't stop thinking about the rocky coast of New England. Part of the purpose of his sabbatical was to decide what research to pursue when he returned to the University of Maine in the fall. On his journeys through Chile and South Africa he'd given his talk on lobster neighborhoods at every academic waypoint. At a meeting of five hundred Latin American scientists his presentation had received an appreciative response. The more Bob thought about what had happened, the more he wanted to probe deeper into the American lobster's secrets back home.

"Okay, big guy," Bob said to himself one day, staring out over the wrong side of the Atlantic, "if you really think you know what's going on with lobsters, why don't you prove that you can predict the future?"

PART FOUR

Surviving

10

The Superlobsters

By the time a group of London entrepreneurs settled Jamestown, Virginia, in 1607, the islands along Maine's rocky coast were already overrun with another type of Englishman. Motley bands of convicts, barflies, and kidnappees—anyone destitute enough to be forced into service aboard a long-distance fishing boat—were camping out on the islands and drying cod. In the early 1600s more than three hundred vessels from all over Europe were fishing in Maine waters, crewed by ten thousand men.

The English didn't have the salt their Continental competitors possessed for preserving cod. So instead, they dried their fish on the islands before shipping the catch to Europe. At first these island drying stations were little more than ragtag summer communities. But soon handfuls of these men, perhaps the ones who had the least reason to return home, were choosing to freeze through a winter on a Maine island that was stocked with firewood and fish sticks rather than puke all the way back to England in the hold of a stinking ship. In the spring of 1622, when the Pilgrims in Plymouth were on the brink of starvation, they headed north to beg food from a community of fishermen who were staying fat eating dried cod on Damariscove Island, a slash of land that punctuates the sea five miles off the coast of western Maine.

On a nautical chart, Damariscove Island looks like a vertical hourglass that has been stretched to its two-mile length by the pull of the North and South Poles. Its slim profile is constricted in the middle, as though by a corset. High tide has the effect of drawing the corset tighter, so that only fifty yards of land remain above water at the island's waist. On either side of this waist is a nearly identical cobblestone cove. One faces east, the other west.

Damariscove Island acts like a breakwater out in the ocean. Strange things have drifted onto the island in days gone by. One tale has it that in the seventeenth century, the island's English owner was beheaded and dumped into the sea by Native Americans while he was on the mainland. His body is said to have washed up on the island's shore along with his dog, which had jumped in after him. The island is supposedly still haunted by the ghosts of the captain and his loyal hound.

On a gray November day in 1987, Bob Steneck's graduate student Richard Wahle donned his scuba gear, tipped backward over the gunwale of his skiff, and splashed into Damariscove Island's eastern cove. He was looking for baby lobsters, not ghosts, but the place was so empty it felt haunted. Rick swam along the bottom, combing the rocks. The island was near his lab at the University of Maine, and he'd received a tip from a fellow scientist that baby lobsters had been spotted around the island in the past. The cove looked like the perfect place for lobster young, with its warm shallow water and dense cobble flooring—a mix of stones and small boulders. Rick saw a few older lobsters, but no babies. Wondering if they might be hiding in the crevices between rocks, he picked apart the stony bottom but found nothing. He'd brought along two assistants from the lab, and their searches came up empty too.

The divers surfaced and clambered back aboard the skiff. The day was overcast and eerily calm for November. Rick motored around the island to the western cove and again the divers pulled on their fins. Underwater it was a mirror image of the eastern cove. Rick swam to the bottom and picked up a rock. This time he found a tiny lobster underneath. The creature was only an inch long, and with a flick of its wee tail it disappeared into a crevice. Rick looked under more stones and found more tiny lobsters. Within minutes he knew that he had uncovered the secret for which he'd been searching.

By the mid-1980s, the question of where floating lobster larvae landed on the bottom and transformed themselves into fully formed lobsters had become one of most maddening mysteries of lobster science. Four hundred miles northeast of where Rick was diving, in a cove on Isle de la Madeleine in the Gulf of St. Lawrence, a Canadian biologist had recently discovered a concentration of baby lobsters that suggested an answer, and Rick believed he was looking at something similar—a lobster nursery. What Rick didn't know yet was that he hadn't stumbled onto just any nursery. The western cove of Damariscove Island contained possibly the highest concentration of baby lobsters in the world.

Rick had first discovered that treasures lay beneath the waves when he was ten. His family was snorkeling in the British Virgin Islands when a man introduced himself as Her Majesty's Salvager. The next day Rick and his parents found themselves motoring out to sea aboard the man's boat. After they had lost sight of land the man shut off the engine and threw over an anchor. While Rick's parents watched, Her Majesty's Salvager hoisted a scuba tank onto their son's small back and told them not to worry, the tank's lack of a pressure gauge was no cause for alarm.

Treading water on the surface, Rick hyperventilated. Her Majesty's Salvager calmed him, and a minute later Rick was submerged in thirty feet of emerald sea, rays of sunlight sparkling around him like strings of gemstones. He hovered over the site of a shipwreck. Tropical fish darted past remnants of a hull and rigging. Cannon lay on the bottom, encrusted with coral.

After graduating from the University of New Hampshire in 1977 Rick took a job monitoring marine life around the site of the Seabrook nuclear power plant on the New Hampshire coast. Underwater Rick counted lobsters among the creatures he studied, but paid them little heed. He spent hours engaged in such tasks as sucking up algae and worms for later study using an underwater vacuum cleaner. Though less romantic than exploring Caribbean shipwrecks, the underwater world off the New England coast captivated Rick just the same. To find terrain on land that was equally free of human habitation required a journey by car of several hours. With a scuba tank Rick could escape from humanity in minutes.

Rick next earned a master's degree studying shrimp in San Francisco Bay. He returned east in 1982, took a job in a natural-history museum, and got married. On a trip to the University of Maine he met Bob Steneck.

What Rick liked about Bob's approach to ecology was the way Bob saw the world not in minutes, hours, or even days, but in hundreds, thousands, and millions of years. When Bob talked about organisms interacting in the wild, whether it was sea urchins consuming seaweed or lobsters fighting for shelter, he always asked what evolutionary process lay behind the behavior. During his research in the Caribbean, Bob hadn't been content to observe the relationship between coralline algae and the snails that ate it. He'd also drilled holes into the reefs to see what the algae had looked like in the past. Then he'd sorted through museum fossils to reconstruct the algal forms and associated mollusk species that had grazed on the algae millions of years before that. Rick enrolled at the University of Maine as Bob's first Ph.D. student in 1985. Rick learned to chant Bob's mantra of ecology—patterns, processes, mechanisms—and turned his attention to the American lobster.

From Bob's censuses of lobster neighborhoods, Rick knew that the number of lobsters in any given location might be limited by the number of hiding places in the rocks. Immersing himself in the latest scientific literature, Rick learned that other researchers had recently proposed similar ideas.

In Woods Hole, Massachusetts, a pair of biologists at the National Marine Fisheries Service had analyzed the data Canadian researchers had collected off Prince Edward Island in the 1950s and 1960s. The data indicated that high numbers of larvae didn't necessarily result in high numbers of lobsters, and the new analysis had concluded that a fixed amount of shelter space might cap the entire lobster population at a fixed upper limit, year after year. In Italy, a fisheries expert at the United Nations had proposed that the fractal geometry typical of the natural world—many small spaces, fewer large ones—could even be used to predict, mathematically, the degree to which high numbers of small lobsters would be limited by the number of appropriate hiding places as they grew bigger, resulting in lower numbers of large lobsters.

In the study of population dynamics there was a term for this sort of limiting factor—a demographic bottleneck—and Rick discussed it with Bob. No matter how many offspring a population produced, the young would have to squeeze through the bottleneck to reach adulthood, and only a certain number could fit. For whatever reason, excess young would die off. Lobster abundance could fluctuate below this limit, but couldn't rise above it.

Scientists in New England, like those in Canada, knew that lobster larvae floated off the coast during the summer and that, a few years later, young lobsters showed up in fishermen's traps. What happened in between was the mystery. If it was the limited number of shelters that was causing a demographic bottleneck, Rick suspected the bottleneck occurred sometime during this cryptic period of early life. Over millions of years, evolution had clearly favored lobsters that fought aggressively to secure protective cover. But when, exactly, in the young lobster's life did shelter matter most, and why?

~

The process of shedding its shell so dominates the life of a lobster that not until it has been living for ten or fifteen years does the interval between molts lengthen enough for the animal to enjoy a period of uninterrupted existence, when it is neither preparing to shed nor recovering from its most recent shed. But at least for a lobster proper, living a rockbound life on the bottom of the ocean, the new shell will always look like the old one. That can't be said of an infant lobster before it settles down.

As an embryo, the lobster begins life as a speck in the mass of dark green yolk inside the egg. Gradually the embryo metamorphoses into what, under a microscope, comes to resemble a translucent louse wrapped in a ball of orange cellophane, topped with an oversize pair of eyes. This period of transformation alone requires the embryo to shed its tiny shell some thirty-five times while still within the egg, developing an increasingly complex body plan with every molt. After hatching, the larva molts through yet another succession of body forms that look like miniature shrimp with spikes. Not until the creature is ready to settle on the ocean bottom does it look like a lobster.

The lobster larva floats near the surface for several weeks. Attached to its legs are paddlelike appendages that make for snazzy acrobatics—mostly somersaults—but aren't much use for covering distance. Then, between its existence as a floating larva and its life as a fully formed bottom dweller, the animal undergoes an identity crisis. It molts to what biologists have termed a "postlarva." The postlarva is about three-quarters of an inch long and possesses the exact appearance of a miniature walking lobster. For the postlarva, however, walking is much too slow.

A fully formed, bottom-dwelling lobster can fling itself backward with rapid contractions of its abdominal muscles, but this emergency escape mechanism is so lacking in grace and control that it can hardly be considered real swimming. By contrast, the tiny postlarva swims nearly as well as a fish, propelling itself forward through the water on a reliable trajectory and at a constant speed by beating the swimmerets under its tail. This is the only time in the lobster's life when the animal can swim forward, and even though the minuscule postlarva is smaller than a baby lobster, scientists have affectionately bestowed it with a grand nickname—the "superlobster." Like a miniature Superman, the animal flies through the water with its little claws outstretched. Its adventures are all the more heroic for being brief. Life as a superlobster seldom lasts more than two weeks. Soon it must settle to the bottom and undergo its final shape shift by shedding its shell yet again. What emerges is an inch-long baby lobster that has lost the ability to swim.

The swimming skills of superlobsters were noted in the scientific literature as early as the turn of the century. But it was Stanley Cobb, the pioneering investigator of lobster behavior at the University of Rhode Island, who demonstrated that superlobsters swim for a reason.

As a graduate student in the 1960s, Stan started with an experiment involving twenty-four plastic dishpans and 120 superlobsters. Six of the dishpans he left bare. The remaining pans he outfitted with bottoms of either mud, sand, or coarse gravel. In the latter case, the gravel provided crevices that a superlobster could tuck itself into. Stan positioned each dishpan under a separate supply of running seawater and dropped five superlobsters into each pan. He fed them a daily dose of brine shrimp and watched to see which superlobsters would settle to the bottom first.

Each superlobster swam around exploring its dishpan for more than a week before any committed to becoming lobsters. But when the settling and molting began, the superlobsters that could hide themselves in nooks between chunks of gravel were, on average, two days ahead of those in the sand pans, who busied themselves trying to dig a depression first. The superlobsters in mud, and in dishpans with bare bottoms, lacked any way to hide and were the last to become lobsters. Clearly, locating shelter was a priority for superlobsters. Without a hiding place they would delay settling to the bottom. Stan thought they must be holding out in the hope of finding more protective terrain.

Watching superlobsters in dishpans was one thing. Trying to track them through the ocean was quite another. That didn't stop Stan. Once he'd finished his Ph.D. and secured a teaching post, he talked a handful of colleagues into spending the month of July snorkeling in Buzzards Bay searching for swimming postlarvae—creatures about the size of kidney beans. In shallow water the divers actually spotted a few superlobsters. To augment their observations they also released seventy lab-reared superlobsters of varying ages. Some were just a day or two old; others had been superlobsters for more than a week.

The scientists discovered that superlobsters swam in two different modes—"claws together" and "claws apart." The latter was slower but useful for hunting at the surface. Claws-apart superlobsters frequently paused to attack foam bubbles floating on top of the water. The behavior made sense once the scientists witnessed a couple of superlobsters carrying winged insects they'd dragged under with their claws. Other superlobsters were seen making a meal of floating crab larvae. When a superlobster near the surface encountered a less appetizing object—a weed or twig, for instance—it stopped methodically, backed up, circumnavigated the obstruction, and resumed travel along its original course.

In claws-together mode, the superlobsters became torpedoes. Streamlining their bodies, they dove into deeper water, sometimes cruising horizontally a foot beneath the surface, other times making a beeline for the bottom. The younger superlobsters, having been in the postlarval stage only a few days, were eager explorers. After swimming like fish for several yards they would dart to the bottom and scurry around on foot, investigating places to hide. As often as not—especially if the bottom was featureless—they would clap their outstretched claws together and, like Superman leaping over a tall building, launch themselves back up so they could fly to a new location.

The older superlobsters, however, were less inclined to leave the bottom once they'd landed, especially if they'd found places to hide. They seemed to hear the ticking of their biological clock. After a larval existence of somersaulting and a postlarval existence of flying, the freedom of weightlessness was now giving way to the harsh realities of growing into a proper baby—gravity, and the need for protection.

Back in the lab, one of Stan's graduate students built a seawater racetrack and recorded superlobsters swimming at speeds of nearly twenty centimeters per second. The superlobsters did their fastest swimming during daylight, in contrast to the nocturnal habits of grown lobsters. Stan and his student guessed that superlobsters might use the sun for navigation—many arthropods tap into celestial cues to establish direction, and some shrimps have been shown to orient themselves in relation to planes of polarized sunlight in the water. The scientists calculated that in five days, a superlobster could swim nearly twenty-five miles. But where, exactly, were they going?

In the western cove of Damariscove Island on that gray November day, Rick Wahle had lifted more stones off the bottom and discovered more tiny lobsters, but he'd also discovered a problem. When he disturbed the stones a cloud of sediment obscured his view. It was obvious that the cove was a nursery for lobster babies, but with clouds of silt everywhere, it was difficult for Rick to get an accurate estimate of their numbers.

Recalling his days vacuuming up algae and worms near the Seabrook nuclear power station nearly a decade earlier, Rick returned to Bob Steneck's lab and rummaged through a pile of unused sections of PVC pipe, but none were quite right. A trip to the hardware store secured what Rick wanted—a section of wide plastic pipe three feet long, along with several sheets of flexible window screening. Rick drilled a hole near one end of the pipe and attached a flexible hose, running the other end of the hose to an old scuba regulator on a tank of compressed air. He folded the window screening into a bag and strapped it over the other end of the pipe, then stood back to admire his handiwork. It looked like a makeshift bazooka.

The following summer Rick and an intern piloted their skiff back out to Damariscove Island's western cove, this time carrying Rick's underwater vacuuming device. They slipped the contraption into the water and swam with it down to the cobblestone floor fifteen feet below the surface. Similar devices—called suction samplers or airlift samplers—were commonly used to collect things like clams as well as algae and worms, and using one to collect baby lobsters seemed like a reasonable idea. But Rick wasn't sure it would work.

Holding the pipe upright, Rick twisted open the valve on the tank. A stream of bubbles shot through the hole and rose the length of the pipe before bursting out through the mesh bag at the top. Rick gave his intern a thumbs-up, then lowered the pipe while the intern lifted stones out of the way. The stream of air generated a suction that inhaled everything into the pipe in a blur. When Rick closed the regulator valve and detached the mesh bag, the sand and silt had mostly blown through the mesh and away from the divers. What remained inside was a collection of shell hash, pebbles, and baby lobsters.

After making some modifications to his suction sampler, Rick was ready to put it to use. He first searched for other nurseries over a ten-mile swath of coastal sea. Between Damariscove Island, which lay exposed five miles offshore, and Pemaquid Harbor, an estuary five miles inside a narrow bay, Rick chose five locations. Each contained three kinds of underwater terrain in close proximity—open bedrock, sediment, and cobble.

Rick had read Stan Cobb's papers on superlobsters—the little postlarvae that swam to the bottom and transformed themselves into the babies Rick was studying. If Maine's superlobsters were like the ones Stan had studied in dishpans, they would avoid both the open bedrock and the less protective sediment. Instead, they would seek out hiding places among the nooks and crannies in the cobble—and there they would stay, molting to become babies. With assistance from Bob and a team of student divers, Rick set out in the summer of 1988 to fulfill his role as an ecologist by mapping the patterns of distribution and abundance of the natural world.

At each site, Rick and his divers began by throwing down a square frame made of rebar, a half meter on a side and spray-painted orange, called a quadrat. By delineating random pieces of bottom that were always exactly the same size, the quadrat helped the divers generate statistically useful data from which they could extrapolate.

When working on cobble, the divers removed the stones inside the quadrat while suctioning away the silt underneath. On mud bottom, the divers suctioned six inches of surface layer away because baby lobsters had been known to drill U-shaped tubes when the mud was deep enough. Sand was too loose for burrowing, so there the divers didn't bother with suctioning but compiled their censuses by sight, as they did for solid bedrock. In a matter of weeks, Rick and his team had sampled more than 350 quadrats and blown through some 120 tanks of air.

The pattern was astonishingly clear. Superlobsters settled, and baby lobsters lived, almost exclusively in shallow coastal seafloor where more than two-thirds of the bottom was composed of cobblestones at least several inches in diameter. Bedrock and sediment were mostly bare of babies.

But the seafloor along the Maine coast was mostly bedrock and sediment. Rick calculated that only about 10 percent of the bottom in his study area contained the kind of prime cobble that baby lobsters required. If the rest of the Maine coast was similar, then the terrain of the seafloor itself might well be the bottleneck that limited the lobster population.

That made the bottleneck almost impossibly tight. Yet there was evidence that something made it even tighter. Even when prime cobble was present it could lie barren. Why, for instance, on the eerie island of Damariscove, was the cobble cove facing west home to more babies than anywhere else, while the cobble cove facing east was home to little but ghosts?

~

Catching insects with a butterfly net hanging off a helicopter might have been easier. Lewis Incze, an oceanographer at the Bigelow Laboratory for Ocean Sciences, was an expert at catching the larvae of crustaceans, and he'd offered to use his remarkable skills to help Rick Wahle unravel the mystery of Damariscove Island. While Rick counted babies on the bottom, Lew searched for superlobsters on the surface.

Lew stole a glance at the sea ahead, then returned his gaze to the net skimming along the water from a boom off the side of the boat. The tiny holes in the net's one-millimeter mesh barely let through water, let alone superlobsters, and Lew had to make sure a floating log or tangle of seaweed didn't tear the net. A rectangular frame at the mouth of the net was attached to a pair of miniature water skis, which kept the lower edge of the opening at a depth of half a meter. If there were superlobsters swimming at the surface, Lew's net ought to snag them.

The boat was running west in a line straight from Damariscove Island. At the end of the tow Lew hauled in the net and dumped out the contents into a sorting tray. A research assistant helped Lew comb through the pile of debris, seaweed, and plankton while the boat's captain steered toward the beginning of the next tow. With the height of the superlobster season upon them—it was the middle of August—the researchers were repeating this routine sixty to seventy times a day.

Lew liked being on the water. He was an avid sailor, and his office at the Bigelow Laboratory in Boothbay Harbor was just a twenty-minute cruise from Damariscove Island. Lew had grown up in Maine and from the age of five had spent his summers living on a tiny island with a single cabin. His graduate work had taken him to the opposite edge of the continent, where he'd earned a Ph.D. chasing the larvae of Alaskan snow crabs through the Arctic climes of the Bering Sea. He was glad to be back in his old neighborhood, because the Gulf of Maine was a unique natural laboratory for the study of oceanography. Whether Lew was at work on a research cruise or enjoying a summer sail, the gulf's constant movement intrigued him—its cold deep currents, its warm shallow eddies, its turbulent upwellings, its prevailing winds. The question of how the physics of the sea interacted with the biology of its creatures fascinated him.

After the last tow on the western side of Damariscove Island the captain gunned the boat for the mile-long trip around the island to the eastern side. The afternoon sun had kicked up a breeze out of the southwest, a trademark of the

Maine coast in summer, and the boat wobbled on the fluttering sea. Lew had yet to analyze the numbers, but so far he'd seen an obvious pattern. On his offshore tows, Lew had been catching similar numbers of superlobsters whether he was on the east or west side of the island. That wasn't the case near shore. When he trolled over the western cobblestone cove where Rick's nursery lay, Lew caught superlobsters in numbers that often exceeded his catches offshore. But above the cobblestone cove on the eastern side, Lew's net usually came up empty.

To a sailor like Lew, who'd spent his boyhood on the coast, the puzzle wasn't difficult to solve. Lew compared his data with wind readings from a weather buoy anchored near Damariscove and saw that the direction of the summer breeze easily accounted for the patterns he and Rick observed. Damariscove Island's narrow, north-south profile ran nearly perpendicular to Maine's east-west coastline. The southwesterly wind was pushing the surface water containing superlobsters up against Damariscove's western edge. At the same time, it blew across the island's narrow waist and pushed surface water away from the eastern edge.

Combined with the effects of tidal currents around the island, the southwesterly breeze was creating a superlobster hot spot on one side and a corresponding shadow on the other, despite the presence of hospitable cobble in both places. Most likely, similar effects had been in play when the beheaded ship captain and his ill-fated dog had washed ashore three centuries before.

Given how common the southwesterly breeze was along the Maine coast, it seemed likely that cobble facing east would almost always catch fewer superlobsters than cobble facing west. The effect would be an even narrower demographic bottleneck than could be accounted for by the terrain of the bottom alone. Back at the Darling Marine Center, Rick Wahle chanted Bob Steneck's mantra of ecology to himself—patterns, processes, mechanisms—and contemplated the picture that was emerging.

The pattern was obvious. Baby lobsters were concentrated

in cobble, particularly in coves facing the prevailing southwest breezes of summer. The general processes responsible for that pattern were becoming clear. Water movements delivered postlarvae against western shores, and then the little superlobsters swam to the bottom and, if they found cobble, molted to become bottom-dwelling babies. The mechanisms behind the processes were more specific—the physics of wind-forcing on surface water, for example, and the navigational and shelter-seeking behaviors that were hardwired into the superlobster's nervous system.

But to understand the behavior of superlobsters Rick had to think not in minutes, hours, or days, but, as Bob had taught him, in hundreds, thousands, or even millions of years. The process of evolution had selected some superlobsters over others, favoring the behavioral traits in evidence today. Superlobsters carrying a genetic code that instructed them to find hiding places immediately had survived and passed on their genes. The rest had died. Why?

The ecological training Bob had given Rick had included the unpleasant lesson that evolution was as often a matter of death as of survival. It wasn't long before Rick discovered the gruesome fate awaiting baby lobsters that never found the nursery grounds.

11

Attack of the Killer Fish

Warren Fernald gaffed his next white-and-yellow buoy and tapped the throttle of the *Mother Ann* down to idle. He spun the wheel to port so the boat would drift in a circle back toward the trap and take the strain off the rope, then tugged up a few feet of line and flipped the slack over the hauler pulley. He pushed the lever on the bulkhead that powered up the hydraulic hauler and threaded the rope between the spinning steel plates. The hauler whined, tightened the line, and reeled it in.

Warren allowed his eyes to wander over Little Cranberry Island in the distance. Before shifting his gaze back to the rope, Warren glanced over his shoulder to check on his son Dan, who was hauling traps nearby in the *Wind Song*. As a precaution, the Little Cranberry fishermen kept an eye on one another.

Warren had learned the hard way how easily a lobsterman could find himself near death. One winter afternoon he was fishing without a sternman. He threw his last trap of the day overboard when a nail protruding from the trap caught his glove and dragged him to the transom. His boots left the deck and he was about to tumble over the stern when the glove came off. Head and shoulders over the water, he had lain there and watched the white shape of a hand sink out of sight.

Another day Warren had thrown a pair of traps off the rail and gunned his engine, only to see that he'd stepped inside a loop of buoy line. The rope pulled tight around his ankle and dragged him aft, away from his engine controls. He dived to the deck, lodged himself under the stern, and wrapped his arms around the mast of his riding sail while the boat raced out of control. As the boat dragged the two traps through the waves, the rope cut into Warren's ankle. In moments the force would pull him overboard. He let go of the mast with one hand and fumbled in his pocket for his jackknife. Opening the knife with his teeth he sawed frantically at the rope until tendrils of twine splayed with the strain and then, with a pop, finally the rope parted. He'd lost his traps but not his life.

Warren stole another glance at Dan. Every parent worried about the safety of his or her children, and perhaps no one more than a commercial fisherman who counted three fishermen among his sons. Warren had to hope that his boys would never come as close to a grisly death on the water as he had. Given the history of Little Cranberry, though, such hopes might be hard to fulfill. The sons of the island had been dying at sea for nearly two centuries.

Before lobsters were worth trapping, fishermen from Little Cranberry Island perished in pursuit of another catch—the cod. Little Cranberry was probably settled around 1762, but the craze to catch cod didn't arrive on the island until 1803, when a man named Samuel Hadlock anchored his ninety-ton schooner, *Ocean,* in the harbor. Hadlock had encountered a streak of bad luck on Mount Desert Island. His house had burned to the ground, he'd sunk two boats, and his father had just been hanged for murder. Moving to Little Cranberry offered a fresh start.

To restore his fortunes Sam hired a crew of local men and sailed for the Grand Banks, a thousand miles away. Sam's luck improved and he filled his hold with fish. He dried them on the shore of Labrador and sailed east, running French and British blockades to sell his catch in Portugal at premium prices. After a trading run to the Caribbean, Sam returned to Little Cranberry a rich man. He bought a fleet of schooners and transformed the little island into a hub of the New England cod trade. At any one time, six hundred ships might be seen in the waters around the Cranberries and Mount Desert.

But the price of prosperity was steep, for catching cod offshore was dangerous work. Almost every family on Little Cranberry Island sacrificed sons to the deep. Sam in particular paid in proportion to his wealth. By the time he died at the age of eighty-four, three more of his ships had sunk and four of his five sons had died at sea. Cod stocks in the western North Atlantic dwindled, and on Little Cranberry Island, the Hadlock dynasty had waned, making way for a new dynasty built on lobsters. By the late twentieth century, the Fernald family had risen to ascendance.

Aboard the *Mother Ann,* Warren finished cleaning out his trap and shoved it overboard. As the buoy line played out over the stern, he steered his boat toward the white hull of the *Wind Song* to greet his son. When Warren pulled alongside, Dan had just gaffed another buoy and was slipping the rope through the heavy steel housing of his hauler pulley. He wound the rope into the sheaves of the hauler and started it spinning before turning his attention to his father. Warren had gotten a couple of words out of his mouth when Dan's rope went taut. There was a crack like a rifle shot and the pulley housing leaped from its mount and slammed into Dan's jaw, knocking him across the boat and onto the deck, where he lay inert, limbs splayed at awkward angles.

"Dan!" Warren shouted.

A full minute passed without a sign of life. Warren maneuvered the *Mother Ann* toward the side of the *Wind Song* so he could lash the boats together. He saw Dan regain consciousness, his head battered but intact. Another close call.

~

Hunched over his lab bench at the Darling Marine Center, Rick Wahle held another baby lobster on the tip of his thumb. Squinting, he used a pair of calipers to measure the length of the animal's carapace—six millimeters. Rick put down the calipers and lifted a piece of fine nylon thread. He'd tied a loop the size of a pinhead in one end of the line. He dipped the loop into a dish of superglue. A tiny drop of the glue stuck to the thread. After blow-drying the moisture off the lobster's back, Rick touched the drop of glue on the thread to the top of the lobster's carapace. It adhered instantly.

Rick sat back in his chair, took a deep breath, and expelled the air from his lungs. Rolling his head to stretch the muscles in his neck, he wondered how he had ever come up with the idea of putting a baby lobster on a leash, let alone doing it to over two hundred specimens.

Stanley Cobb's experiments had shown that when a superlobster's biological clock ran out of time, it would swim down and molt into a bottom-dwelling baby even when all it could find was open bedrock or bare sediment. If this happened in the wild and the lobsters survived anyway, then Rick's hypothesis that baby lobsters required cobble would be wrong. But if they failed to survive, then Rick would have found additional evidence of the demographic bottleneck. Only baby lobsters hiding in a cobble nursery would live. Those that failed to find cobble would die.

The fine nylon thread and superglue worked only on the smallest lobsters. On his one- and two-year-old babies Rick tied a piece of braided fishing line snug around the animal's carapace like a belt and attached a steel fishing leader. The lobsters would reach down and grab the leader with their claws, but they couldn't break the metal. Rick dangled each animal in the air on its leash and bounced it up and down a few times to ensure that the tether was secure before dropping the baby lobster in its own private dish.

Borrowing Bob Steneck's leaky houseboat, Rick took a summer intern from the lab and moored the boat in the bay. Rick and his assistant splashed overboard, spread patio tiles out at intervals of a few feet, then returned to the boat. Rick had sorted his pretethered baby lobsters into three size categories. With the lobsters in bags, they dove again and clipped each lobster's leash to a tile. Once in place, the lobsters scurried frantically about their individual patios, but like a puppy tied to a stake, each lobster hit the end of its tether before it could run away.

The next day Rick and his assistant dove again. All five of the largest baby lobsters sat calmly on their patio tiles in the sunshine. But a few feet away, while two of the mid-sized babies remained, next to them were three empty loops of fishing line. And on the tiles that had held the smallest babies, all that remained were five wispy strands of nylon, each ending in a tiny chunk of lobster shell still stuck to a ball of superglue.

Rick positioned an underwater video camera next to the patio tiles and ran the video feed to the surface. Over the course of a week, he and his assistant tethered fifty more lobsters to the patio tiles in batches. The scientists fixed mugs of hot chocolate, then sat in front of the TV on the houseboat to watch what happened. A type of small fish called a sculpin would often appear, circle a hapless baby, then clamp its toad-like mouth over the little lobster and shake it until the lobster had disintegrated enough for the fish to tug it free. Another type of small fish called a cunner, with a smaller mouth, tended to strike once, swim away, and return to strike again, the lobster dangling from its tether in successive states of dismemberment.

Each little lobster's death was a heroic sacrifice for the sake of science. The experiment revealed the natural process—fish predation—that enforced the demographic bottleneck. A superlobster that failed to find shelter before transforming itself into a baby would become fast food for these enforcers. Rick's smallest babies were usually attacked by fish within fifteen minutes. The mid-sized babies sometimes lasted an hour. The biggest babies usually remained unscathed. The lobsters in this last category were about the same size as the smallest lobsters that Rick had found wandering across open terrain during his surveys. Having passed successfully through the bottleneck, they had grown large enough not to be attacked. So far, Rick's theory that small lobsters needed shelter was holding water.

The leaky houseboat wasn't. The University of Maine had started to fear for the safety of the students in Bob Steneck's lab, and shortly thereafter the craft was condemned.

~

Jack Merrill had never gotten around to circumnavigating the globe in his sailboat. But in search of lobsters he'd made a habit of running the *Bottom Dollar* twenty miles offshore to a tiny, distant island that had been named—after its big brother, Mount Desert Island—Mount Desert Rock. The remote islet was little more than a half acre of granite protruding from the sea, topped by a stone lighthouse—the easternmost lighthouse in the United States—and a keeper's lodge.

Lobstermen had put a stop to the government's additional increases in the minimum legal size of lobster, and Jack hadn't noticed any dispersal of legal-sized lobsters away from the coast—the consequence Bob Steneck had warned against. But in a sense, Jack was a bit like a large lobster himself. As the number of traps escalated in the shallows, Jack preferred to give the other lobstermen space and seek out new territory in deeper water. His powerful boat gave him that option.

It was an option fraught with danger. The winter storms that ravaged the Rock were so wild that waves had once rolled a seventy-five-ton boulder from one side to the other. The lighthouse keeper and his family had routinely locked themselves inside the tower with a winter's worth of provisions lest they be swept into the sea. To go outside they had tethered themselves to the granite with the stoutest rope at hand. Now the lighthouse was automated and the Rock uninhabited.

Jack squinted at the lighthouse in the gray December morning and guessed it was about half a mile off his stern. He checked the chart he'd spread on the bulkhead and ran his index finger along the contour lines of the underwater terrain. Just southeast of his position the map indicated a special feature on the bottom: "Unexploded ordnance."

Fishing around the Rock was tricky enough without the added risk of hauling up a forgotten bomb or torpedo. Jack gave the boat a drink of diesel and the engine roared. Watching his Fathometer, he circled several alternative locations and found a piece of bottom that looked promising. Despite the risks, setting traps this far from land allowed Jack the luxury of forgetting about the politics of the lobster industry. Out here he could focus on fishing.

"*Bottom Dollar,* you on there, Jack?"

Jack would have recognized that raspy voice crackling over the VHF in his sleep. It belonged to the only other Little Cranberry lobsterman crazy enough to fish this far out. Mark Fernald.

"Yo!" Jack yelled into his radio mike, not bothering to remove it from its overhead clip.

"Kinda slatty out here this morning, isn't it?"

On his chart, Jack finished plotting the path of his next string of traps before glancing again at the waves rolling toward the boat. There were days out here when the waves were so huge they could tip the *Bottom Dollar* nearly on its side—Jack had once lost twenty or thirty lobsters over the gunwale. There were days when he would decide that he wanted to watch his daughter and son grow up. On those days, instead of hauling the rest of his gear, he'd set a course for home.

"Could turn a little nasty," Jack responded, this time tugging the mike off its clip. He scanned the horizon for Mark's boat. "Where you at?"

"Just finished setting some gear out on the ridge."

Inner Schoodic Ridge, an underwater mountain range surrounded by canyons, was ten miles northeast of Jack's position.

"Yup. Well, don't stay out too long," Jack laughed. Mark could drive the dockworkers back at the co-op crazy. Some days he wouldn't pull up to the wharf until the other fishermen were already home eating dinner.

Jack was happy to let Mark fish the ridge. Always interested in biology, Jack liked Mount Desert Rock for the seabirds and other marine life it attracted. As he replaced the microphone, shearwaters were skimming the chop around his boat like stunt planes. A pair of razorbills streaked by. Sometimes gannets circled overhead and plummeted into schools of fish. Porpoises and white-sided dolphins plied the waves, and in summer bluefin tuna leaped from the water in feeding frenzies. And every so often, when Jack hauled up a trap around Mount Desert Rock, the wire cage would break the surface like a time capsule. Inside would be the shimmering bronze body of a cod.

Rick Wahle and Bob Steneck pulled on their dry suits and checked their scuba equipment. They were seventy miles south of Mount Desert Rock, and seventy miles due east of the New Hampshire coast. That put them near the middle of the Gulf of Maine, which was much farther from land than any sane lobster scientist should go.

Rick and Bob were preparing to dive on an underwater mountain range called Cashes Ledge. The slopes on either side dropped six hundred feet and, like Mount Desert Rock, the bottom nearby was littered with live artillery shells from military tests. Ten-foot blue sharks infested the waters, and ocean swells blew like gales across the ridge.

On the deck of the research ship Rick took a couple of deep breaths to calm his nerves. He wasn't sure which was worse, sweating his balls off on the surface in his dry suit or thinking about the near-zero temperatures he'd hit a hundred feet down. The jagged undersea summit they were planning to land on was called Ammen Rock. Rick checked his tank, pulled down his mask, and said a prayer.

Carrying steel bolts two inches thick, balls of epoxy, and bundles of rope, Rick, Bob, and two dive buddies splashed overboard and swam straight for the bottom. Despite the danger of the dive, Bob had to wonder if getting underwater wouldn't turn out to be safer than staying aboard the ship. The cook, the proverbial drunk, had passed out while baking a turkey and had set fire to the galley—he'd doused the bird with a fire extinguisher and served it anyway. Then Bob's female research assistant had discovered a member of the ship's crew stealing underwear from her laundry basket.

Rick, however, was already feeling nostalgic for the antics back aboard the ship. Halfway to the bottom, he could feel his face going numb from the cold and a chill seeping into his arms and legs. The divers had undergone intense training and physical exams, and were breathing a carefully monitored gas mixture called nitrox, which extended their total dive time to about half an hour. But their return ascent would have to be slow so their bodies could reacclimatize, and that left only thirteen minutes of working time on the bottom. Any longer and so much compressed gas would have seeped into their blood that returning to the surface would fill their veins with bubbles, a condition called the bends. Worse, if they waited too long and came up too quickly, the gas could blow holes in their lungs. The former would cause a slow death; the latter a more rapid one. Both were extremely painful.

Rick forgot his fear when he saw the fish. The slopes of Ammen Rock were an undersea park of waving kelp and sea anemones, and over them swarmed vast schools of cod. The scientists felt as though they had dived not just into the sea but into the past, and in a sense they had. Cod continued to thrive on Cashes Ledge because the abrupt rise was murder on commercial fishing nets, which snagged on the sharp peaks. Bob and Rick had come to Cashes Ledge for its resemblance to what the bottom of the entire Gulf of Maine must have looked like before the arrival of modern fishermen—an ocean floor teeming with large predators.

In his predation tests near shore, Rick had been surprised at how soon baby lobsters seemed to outgrow their predators. Lobsters that had reached a body length of about two and a half inches were all but immune to attacks by small fish such as sculpins and cunners. But during most of the American lobster's existence in the Gulf of Maine, the bottom had swarmed with huge codfish. With these large predators in the sea, the risk to young lobsters must have been severe. With its schools of cod, Cashes Ledge gave Rick and Bob the perfect natural laboratory for reconstructing the past. Whether the lobsters they'd brought with them were quite so eager to undergo time travel was another question.

When the divers reached the ridge they went to work molding their balls of epoxy to fit cracks in the rock. Before the glue set, they jammed in a bolt, then tied in a section of rope. Rick and Bob knew that most scuba-diving deaths resulted not from drowning on the bottom but from separation of the diver from the boat. When swells rolled in from the Atlantic and hit the topographical bump of Cashes Ledge, water surged over Ammen Rock at speeds that could drag a person several hundred yards in a few minutes. By installing ropes along the bottom the divers would be within grabbing distance of a handhold while working. A vertical line connected them to the surface so they could raise and lower themselves through the treacherous winds of water, like human flags on a hundred-foot pole.

Soon Rick, Bob, and their assistants had arranged the patio tiles on the bottom. Again, the lobsters were divided into three size classes. But unlike Rick's predation experiments near shore, out here the scientists would be offering up more than just baby lobsters as fish food. The lobsters in the largest category were big enough for a human dinner. Rick and Bob tied the animals to the patio tiles and returned up the rope to the boat.

Twenty-four hours later they dove again. The cod had barely bothered with the babies. Instead, the fish had gone straight for the big lobsters, and they'd obliterated them.

~

The phylum of "jointed-leg" creatures called Arthropoda includes, among other subphyla, the Insecta and the Crustacea. The subphylum of Crustacea is home to fifty-two thousand species of lobsters, crabs, shrimps, crayfish, and their close relatives, but also includes barnacles, sea fleas, pill bugs, and wood lice. The ancestor of the lobster, a shrimplike animal, appears in the fossil record in Ohio and upstate New York as early as the Devonian period, about 400 million years ago. By the time of the Jurassic period, 150 million years ago, clawed lobsters were so similar to today's specimens that if one could be boiled up and served in a restaurant, few diners would notice the difference.

Beginning in the Jurassic, clawed lobsters experienced a hundred-million-year heyday. In the depths of the Tethys Sea, precursor of the Mediterranean and Caribbean Seas, they diversified into at least fifty-three species and rapidly colonized the world's oceans. The lobsters spread west and north along the shores of northern Europe, past Greenland to the continental shelf off Labrador, on to Nova Scotia, and into the Gulf of Maine.

But the lobsters' foray into shallow water exposed them to new risks. Sixty-five million years ago a cataclysm enveloped the earth, obliterating the kingdom of dinosaurs and exterminating many marine species. The clawed lobsters suffered too, and in shallow water their numbers declined. But mass extinctions were soon the least of their worries.

The clawed lobsters rebounded in shallow water until about thirty million years ago, when the fossil record reveals another drastic loss of clawed-lobster species from the continental shelves. This time, however, the eclipse of the lobsters was the result of a family feud. From the basic lobster body plan a new crustacean had evolved that was faster, smarter, meaner, and far more adaptable to the challenges of shallow-water life. Lobsters had given rise to their own toughest competitor—the crabs.

The lobster's armor and weaponry would seem to make the animal impervious. In the seventeenth century, Hungarian warriors wore a type of headgear called the *rakfarkas sisak*—"crayfish-tailed helmet"—because it had overlapping steel plates that curved down to protect the neck, resembling the tail of a crayfish. Soldiers throughout Europe adopted the helmet. Among English cavalrymen it was known as the lobster-tail burgonet. If these warriors had been up on their evolutionary biology, however, they might have chosen a different name.

Compared with a crab, a lobster was vulnerable, and its tail was its greatest liability.

To a bottom-dwelling fish with strong bony jaws, the meaty lobster tail and its crunchy, calcium-rich shell made the perfect meal. At the approach of one of these ferocious fish—perhaps the precursors of the cod—a crab could bury itself in the sediment in a puff of sand. Although a lobster's powerful tail did give it the ability to escape in backward spurts, the fast-fiber muscle tissues of the tail tired quickly, and for a lobster to bulldoze itself a burrow could take several hours. As a result, while crabs ranged widely, lobsters generally limited themselves to neighborhoods with complex, stone- and boulder-filled bottoms, where a few flips of their tails could help speed them into nearby shelter. While the number of crab species exploded into the hundreds and then thousands, the number of lobster species in shallow water declined. Most of the clawed lobsters still living today—about fifty species—returned to their refuges in the deep. Only twelve species of clawed lobster have managed to survive near shore, and the American lobster is the only one to have achieved much success. Given the damage that the codfish, long the most abundant predator in the Gulf of Maine, can inflict on a lobster, that success is exceptional.

~

Bob Steneck was eight minutes into another dive on Cashes Ledge when he lost sight of his dive buddy. Bob crawled forward but didn't see anyone, so he turned around and swam back through the kelp to his rope handhold. Except he couldn't find it. He pushed on, then came to the edge of the kelp and realized he'd gone too far. He was lost on Ammen Rock in a jungle of seaweed.

Bob had logged more than fifteen hundred dives in his career. He'd endured saturation dives as an aquanaut aboard the undersea equivalent of the international space station, the Hydrolab, where researchers lived for a week at a time. But he'd never been this scared before. Surfacing without a lifeline would leave him exposed to the swells that washed over the ridge. He could be dragged so far that his ship would never find him, even if he reached the surface. But he also had less than four minutes of bottom time left before he risked death from the bends. Bob decided to stay low in the seaweed and crawl forward, staying alert to his direction and distance. If he didn't find one of the rope tethers, he could retrace his progress and try another direction. If a swell rolled through, he could grab onto the base of a kelp frond and hope it would hold.

On his first pass Bob became disoriented. He tried to double back, but he could easily have been going farther in the same direction. His heart was pounding. The seaweed was so thick it was hard to see his hands in front of him. He checked his timer. Two minutes remained before he had to begin his ascent.

Bob emerged into an abrupt clearing, but it wasn't a place he recognized. In it lay a ship's anchor, by the look of it abandoned more than a century ago, back in the days when cod had been caught with hook and line. Perhaps the crew, like many of the men of Little Cranberry Island, had died in their quest for cod, and perhaps Bob was the first man to see the anchor in a hundred years. It was encrusted in coral, and its graceful flukes curved toward him like arms beckoning. Perhaps Bob too would die on Cashes Ledge, in his own weird quest for cod.

Mankind's pursuit of the codfish has always been deadly, but it has been under way in the Gulf of Maine for much longer than a few hundred years. The first fishermen in Maine didn't catch cod from schooners equipped with iron anchors—they used dugout canoes and hooks carved from deer bone, as early as 3000 BC. Most of the bones in the rubbish heaps, or middens, of ancient Native Americans are from three- and four-foot-long cod. Maine's natives hauled up a respectable catch of the fish for the next five thousand years.

Strangely, the ancient middens contain no lobster shells. The chitin in crustacean armor decomposes more quickly than bone, but the thick ridges of a decent-sized lobster's claw might have been preserved. Thinner clam shells have lasted. The absence of lobsters is striking because early English settlers found lobsters two and three feet long along the Maine coast and witnessed Native Americans eating them.

What's also striking is that at least some of those English settlers found large codfish hard to come by near shore. In the early 1600s, propaganda aimed at prospective settlers lauded the abundance of large cod in New England, and certainly the fish were plentiful on the offshore banks, where professional fishermen dropped their lines. And yet when settlers had tried to establish a village called Trelawny on the coast of western Maine in 1633, within a decade they were unable to catch enough cod near shore to feed themselves.

On Cashes Ledge, Bob Steneck had seen what a three-foot cod could do to a lobster, and if three-foot cod had been common enough along the coast to be hooked from canoes five thousand years ago, the reason ancient middens didn't contain lobsters might simply be that there hadn't been very many lobsters worth catching—the cod ate them first. Bob guessed that for much of the Gulf of Maine's history, while crabs and other small bottom-dwellers might have flourished, Maine's emblematic crustacean, the lobster, had probably been a marginal resident, struggling to survive the gauntlet of cod mouths. The lucky lobsters that made it through the gauntlet did so because they outgrew their predators.

In those days, the demographic bottleneck that controlled the size of the lobster population might not have been limited to the nurseries. Another bottleneck would have occurred later, when young lobsters became a meaty-enough meal to attract the attention of a sizable cod. The patterns, processes, and mechanisms of lobster ecology would have been different, and the demographic shape of the lobster population unrecognizable. If Bob and Rick could have donned their scuba gear five thousand years ago, what they might have found was some tiny baby lobsters and some huge old lobsters. But lobsters in between, the young adults that today wander freely and supply a vibrant fishery, would have been attacked by cod the instant they were exposed in the open.

By the time European settlers arrived in the middle of the

second millennium, however, cod populations near shore could have been thinned somewhat by thousands of years of Native American fishing. Indeed, recent scholarship suggests that Native American civilizations had a greater impact on the environment than was previously realized. What's more, many scientists believe that beginning around the year 1400 the Northern Hemisphere was chilled by the Little Ice Age, which lasted nearly half a millennium, and could have pushed some of the cod in shallow waters away from the coast, perhaps accounting for the scarcity at Trelawny. If so, more lobsters might have grown to the proportions of invincibility, accounting for the large specimens commonly noted by early explorers. That these monstrous lobsters could be found in shallow water fit with Bob's understanding of how lobsters competed for shelters. Without younger lobsters filling up coastal habitat, the big animals didn't feel pressured to leave for deeper water, as they seemed to today.

By the second half of the 1800s, men in dories in shallow water were trapping enough of these big lobsters to create a new crisis. Lobsters were being eaten by two predators now—the cod that remained, and humans. That, Bob suspected, had been the one-two punch the lobster population couldn't endure, causing it to crash in the early twentieth century.

Then the lobster's fortunes had reversed again. In the 1930s and 1940s cod were dragged toward commercial extinction by the invention of the modern trawler. With its oldest predator on the run, the American lobster was finally given free rein along the Maine coast, probably for the first time in its history. To the south—off Rhode Island, Connecticut, and New York—wide-ranging fish from the Atlantic Ocean and the tropics still attacked lobsters. But inside the semienclosed Gulf of Maine, thanks to human fishing, clawed lobsters were probably handed their best opportunity to expand in thirty million years.

Now, over a few generations, Maine's fishermen had switched lifestyles. Instead of hunting scarce cod offshore they had settled down to farm their local lobster plots, setting rows of traps almost as though they were cultivating corn. In the vocabulary of ecology there is a term for this type of human activity: "fishing down the food web." With the apex predator out of the way, species that are lower on the pyramid explode in abundance and become the new human harvest. It's a nearly universal phenomenon in the sea.

But too much fishing down of the food web can destroy an ecosystem. The government scientists charged with protecting America's marine resources—or rather, what's left of them—have long worried that after the cod, the lobster will be the next species to be decimated by overfishing. They cite the decline of large lobsters near the coast since the 1800s as a sign of the lobster's impending demise. Lobstermen have reseeded their crop by V-notching, and smaller lobsters have exploded in abundance. The result, however, is a population structure that has never before been tested in nature. This makes many observers nervous. There's little comfort in the irony that the lobster population's great expansion in the twentieth century was partly due to overfishing. Today cod have all but vanished—so much so that one of the few places they remain is Cashes Ledge.

For all the splendor of its plentiful codfish, Cashes Ledge was not a place where Bob Steneck wanted to die. Underwater, he surveyed the clearing again and detected a slight slope to the rock, heading upward in the direction from which he had come. If he could feel his way back upslope, there was a chance it would take him back toward his original position. He plunged back into the kelp. Surrounded by a waving forest of seaweed, he swam forward and again checked his timer. Thirty seconds of bottom time remained. He could surrender and stay here, with his feet touching the earth, or he could launch himself into the sea.

Praying that the ocean would remain still, Bob aimed himself upward and began to kick slowly with his fins, trying to maintain the trajectory he thought might take him toward the boat. Suddenly a bubble of air floated up a few yards in front of his face mask. Then another. He waited. A third bubble rose, this time off to his right. He adjusted his course and saw more bubbles ahead. Bob guessed that by an enormous stroke of luck, he was above his dive buddy, who would have just started to surface too. Bob hoped the man was on the tether. If he was, Bob could follow the bubbles and intercept the rope. All he needed was for the ocean currents to stay quiet for another few minutes. They did, and moments later Bob saw the line leading to the surface.

12

Kindergarten Cops

The R/V *Argo Maine* was chugging up the coast at a steady nine knots, her squat white hull ready to tackle an oncoming swell. But she wasn't known for riding waves with grace. Originally operated by Oregon State University, she'd been called the *Cayuse*. In a local Native American dialect the name meant "wild horse," but "bucking bronco" might have been more apt. When the Maine Maritime Academy took over the boat she quickly earned the nickname "Vomit Comet."

Bob Steneck and Rick Wahle had commandeered the aging eighty-foot Vomit Comet for a series of scientific commando raids along the coast, between northern Massachusetts and Canada. The scientists wouldn't be confronting deadly offshore currents on this trip. They had a team of student divers aboard, and a geologist had come along from the University of Maine with a sophisticated sonar system, which he would use to locate fields of cobblestones in shallow water.

The larvae specialist Lew Incze was aboard a smaller research boat, his net for catching superlobsters stowed on deck. Come nightfall, Lew would stow himself on deck too. Most of the team would sleep in cabins inside the *Argo*'s hull, but Lew had brought a tent to pitch next to his equipment on the smaller craft. He snored so loudly it was the only way the others would get any rest.

The Vomit Comet steamed past the hills of Mount Desert Island and motored Down East to a remote coastal town called Jonesport, fifty miles short of the Canadian border. The northern air was cool and a dense bank of fog hugged the rocky coast. The geologist deployed what looked like a torpedo into the water—a "sonar fish" for mapping undersea terrain—and strafed the bottom with acoustic signals. Bob, Rick, and their assistants hustled into scuba gear and dropped into a skiff powered by a sputtering outboard motor.

"You see that isolated knob?" the geologist shouted through the mist. "Closer in. One, two hundred meters due north, that way."

Bob nodded and gunned the outboard. By now Bob and Rick considered themselves rather skilled at finding baby lobsters. They adjusted their face masks and regulators with confidence and splashed overboard.

The purpose of the trip was to search for additional lobster nurseries in new parts of the coast. In the western half of coastal Maine where they'd conducted most of their research so far, Rick and Lew had seen that baby lobsters were limited to certain cobble plots. They'd also seen that once a baby had spent its first year or two hiding in a cobble plot, it could emerge and forage in the open without fear. With large predators confined to places like Cashes Ledge, a baby lobster safe in a nursery had already passed the most difficult test of its life—that is, until it reached the minimum size for harvest. The next step for the scientists was to see if similar cobble plots were populated with baby lobsters in other parts of the coast.

The genius of Rick's underwater vacuuming system wasn't simply that it could help identify nursery grounds, but that it could be used to sample population densities of baby lobsters reliably, and with repeatable techniques. Using the protocols and equipment Rick had developed, anyone could generate statistics on the abundance of new baby lobsters from one year to the next. Rick and his colleagues envisioned an annual index for lobster settlement along the entire coast of Maine, a surveillance system for monitoring nurseries and thus the future of the lobster fishery. Perhaps someday there would even be a settlement index that could predict the future for all of New England.

Ideally, having such a system would give fishermen and

government scientists a new tool for assessing the health of the lobster population. By analogy, the health of the U.S. economy was judged using a variety of indexes—retail sales, manufacturing inventories, employment rates, housing starts, and so on. Similarly, the state of the lobster stock could be judged not just by catch statistics and estimates of fishing effort, but by settlement surveys as well.

The divers sank into the water off Jonesport fully expecting to locate baby lobsters on the bottom. Once their eyes adjusted to the dim light they found themselves hovering over a perfect nursery—just the sort of cobble that baby lobsters required. The divers saw some impressive adult lobsters, larger than what they were used to along the western half of the Maine coast. But after a long day of diving across several acres of terrain, the scientists were stunned. They had not located a single baby.

The *Argo* backtracked west to the waters off Mount Desert Island and anchored a few miles from Little Cranberry. The next morning, Jack Merrill arrived in a lobster boat with several other fishermen and boarded the *Argo* to watch the scientists work.

Again the divers suited up and dropped overboard, and again they were spellbound by the expanse of perfect cobble. Acres of glacial moraine, deposited during the last ice age, formed swaths of nursery habitat more expansive than what Rick had seen in western Maine. Again, they noticed adult lobsters roaming among the rocks. But after hours of diving the scientists remained foiled. Babies were nowhere to be found.

"Hey boys," Barb Fernald called, "don't forget your helmets!"

On Little Cranberry Island, Bruce and Barb had bought bicycles and matching helmets for their twin sons. The boys had just finished kindergarten. By fall, they would be accustomed to riding down the island's main street to the two-room schoolhouse.

"We already got them on, Mom!"

"Good job. Now let me see."

The boys raced into the kitchen and stood at attention side by side. Barb had to suppress a laugh. The helmets strapped on top of their tiny bodies made them look like a pair of human lollipops. Barb pulled them to her in a hug and then sent them on their way. As she watched them jump on their bikes and pedal off toward the village, she knew the other islanders would help keep an eye on them. Some folks still had trouble telling the boys apart. The giant helmets covering their heads wouldn't help.

Yet in time, her boys would discover their own identities. As a young woman Barb had struck out on her own and built herself an unconventional life. She didn't regret it, although the path she'd chosen hadn't always been smooth. After Barb had quit lobstering with Bruce she had suffered from alcoholism during the long days on land. By the time her sons were three she'd been ready to seek help. Just getting to the Alcoholics Anonymous meetings had required her to arrange child care and transport from Little Cranberry to the mainland. Avoiding drink remained a daily battle, but now, with the twins in school, Barb had been sober for three years. Fighting her disease had helped her reaffirm her love for her family and the island life she had chosen.

When Barb was a little girl, her parents could not have guessed that she would become a commercial fisherman and a year-round resident of a small island. Barb wanted to give her own children the chance to do something novel with their lives. She and Bruce had talked it over. He was the fifth generation of Fernalds to be a fisherman on Little Cranberry, and the boys were free to become fishermen too. But Bruce hadn't gone to college, and he wanted his sons to have that opportunity—and the chance to try another occupation besides fishing. Bruce and Barb would have to start saving money for simultaneous tuition payments, and hope that the lobster catch over the coming years stayed strong. The fact that the waters around Little Cranberry were empty of lobster babies hadn't factored into their calculations.

Soon it was midsummer and catches slowed as the lobsters went into hiding for the shed. Bruce woke one morning at the leisurely hour of seven o'clock and motored the *Double Trouble* over to the yacht yard on Mount Desert. Waiting for him was the Travel Lift, a framework of blue girders the size of a house, riding on wheels from a 747 jumbo jet. The straps tightened under the *Double Trouble*'s hull and raised all eight tons of her from the ocean like a whale in a gurney. Her dripping hull was spotted with algae.

Watching his boat—his entire livelihood—sway in midair on its way down the street was a disconcerting experience. Bruce was relieved when, fifteen minutes later, the *Double Trouble* was resting on blocks in the parking lot. Over the next three days he and his sternman scrubbed the grime and herring stains off the cabin interior and repainted the red trim. Above the waterline they buffed and waxed the hull to a glossy white finish, and below they slathered on a fresh coat of antifouling paint to keep the algae at bay for another year. Finally, from a receptacle above the rudder Bruce removed a pocked chunk of zinc and replaced it with a fresh slab. Zinc surrendered its electrons more readily than steel and would serve as a sacrificial offering to the sea, distracting the salt water's corrosive powers from the functional metal parts that protruded from the boat's hull.

Shiny and shipshape, the *Double Trouble* settled into the water, and Bruce steered her back to the island. After tying up his skiff at the co-op Bruce strolled over to the bar on the restaurant wharf and bought himself a beer.

By August Bruce had little time for such indulgences. He'd set three hundred traps in the shallows around the island and another five hundred on the plateaus and gullies leading into deeper water, and he was working ten-hour days. The catches were decent enough, but not huge.

At the end of the summer the islanders gathered in the Grange hall for a traditional evening of skits. Parents took their seats in rows of folding chairs while their children scurried into position backstage. Bruce was exhausted from another day on the water and hoped he could stay awake.

The lights dimmed and the island's children launched into a hastily rehearsed rendition of "The Tortoise and the Hare." At the climax of the race, the hare zipped around with such speed that it seemed to be coming from both sides of the stage at once. Then the audience realized that in fact it was, and they roared. Bruce guffawed, and Barb, her chin in her palm, broke into a grin. Their twins, paper rabbit ears taped to their bicycle helmets, were streaking back and forth, impersonating each other and laughing themselves silly.

The twins quickly outgrew their helmets and soon wanted bigger bikes, and by the following summer the *Double Trouble* was again in need of maintenance. Bruce applied a new coat of antifouling paint and replaced the zinc. But this time, when he returned the boat to the water and began hauling his traps, something in the lobster fishery had changed.

"*Double Trouble.*"

It was his brother Mark calling on the VHF. Bruce's next trap broke the surface, and he pulled it onto the rail before answering. Mark had recently bought a big new boat. The boat's gray hull was supposed to have carried the name of his wife, but he'd just gotten divorced. Instead, he'd dedicated it to his three-year-old son.

"Go ahead, *Merry Marcus*."

"I talked to a fellow to the west'ard there the other day," Mark reported. He was referring to one of his friends who lobstered in an area about fifty miles west of where Bruce and Mark hauled their traps. "He said he had twelve hundred pounds in one day."

"Jesus," Bruce said, "guess he can retire early."

"I know it," Mark said.

The Cranberry fishermen had heard that catches to the west were suddenly on the rise, but twelve hundred pounds was an extraordinary haul. Bruce wondered where those lobsters had come from, and why he and his brothers weren't seeing a similar increase Down East.

A few years had passed since the scientific commando raids Bob Steneck, Rick Wahle, and Lew Incze had launched from the Vomit Comet. After completing his Ph.D. at the University of Maine in 1990, Rick had landed a postdoc at Brown University and then a lecturing post at the University of Rhode Island. But that hadn't stopped him from returning to his baby lobsters in Maine at every opportunity.

Rick and Lew had selected eight nursery sites in the region of western coastal Maine where Rick had conducted his original surveys, and begun counting swimming superlobsters and baby lobsters on the bottom every summer. Rick had also found nurseries in Rhode Island Sound and was monitoring them with his vacuum cleaner. Scientists in Massachusetts were considering a similar program. The lobster-settlement index that Rick and his colleagues had envisioned was gaining acceptance.

But a crucial pair of questions had arisen. Why had the ecologists found lobster babies only along the western half of the Maine coast? And why had lobster catches suddenly increased, but also only in the western half of the state? By the early 1990s the geographic pattern had become obvious. The densely populated nurseries that Rick and Lew monitored coincided with rising catches in western Maine. The barren nurseries that the ecologists had discovered Down East coincided with stagnating catches in eastern Maine. The scientists wondered what accounted for the difference between east and west, both in the nurseries and in catches.

On that day Down East aboard the Vomit Comet, when Jack Merrill had joined the scientists in the waters off Mount Desert Island, the divers had found not a single baby on the bottom. But while they were diving, Lew had been towing his superlobster net along the surface. To his surprise, Lew had hauled in a thick run of swimming superlobsters.

The ecologists concluded that those superlobsters had been on a sort of suicide mission. The cobble below was a maze of perfect hiding places, yet the superlobsters weren't landing.

Almost certainly they had ended up dying at sea. It was possible that the water below the surface had been too cold for them to settle. And if cold water could leave perfect nurseries in eastern Maine barren, maybe warm water was responsible for the densely populated nurseries in western Maine.

It seemed like a reasonable hypothesis. Alternative explanations for the sudden rise in catches in western Maine weren't convincing. In the cod fishery, catches had increased as a result of new technology—nineteenth-century cod fishermen had dropped baited hooks from dories, but by the 1950s, ships the size of factories were consuming entire schools of cod with giant trawl nets and freezing the fish at sea. No similar technological revolution had occurred in the lobster fishery. Except for the switch from wood to wire, the lobster trap had changed little in a hundred years. Even quadrupling the number of traps in the 1970s and 1980s hadn't resulted in the capture of more lobsters. Instead of blaming the increase in catches in the 1990s on better trapping techniques, the ecologists were forced to entertain the mystifying possibility that the lobster population had simply grown in size—the exact opposite of what government scientists had warned would happen.

Nor was this increase likely to have been caused by the recent decimation of the codfish, the lobster's most deadly predator. In coastal waters near shore, the destruction of cod stocks had occurred much earlier in the twentieth century. This time something else was going on.

Lastly, the government's partial increase in the minimum size was also insufficient to explain the increase. Only a few years had passed, and it took a lobster about seven years to grow to harvestable size.

Perhaps the lobstermen's own conservation techniques—V-notching and the maximum-size law—had been effective in ensuring a surplus of eggs all along. Even if they had, however, dumping more seed on a field—as Bob liked to point out—didn't necessarily increase the yield of crops. Bob, Rick, and Lew had to conclude that the catches were on the rise because the neck of the demographic bottle had somehow widened.

More little lobsters were getting through, perhaps because of recent changes in water temperature.

Once again, Bob was reminded that populations of organisms were always in flux. By 1994 he was eager to return to Mount Desert Island to take another look. If the bottleneck was widening, perhaps this time Bob would actually find some baby lobsters in the eastern half of the state. And at the very least, if he and Rick could map the pattern of baby, juvenile, and adult lobster abundance along the entire coast as catches rose, they might be able to identify more accurately the process at work.

Meanwhile, another scientist had found even more baby lobsters in the western half of the state, and she'd done so without even trying.

~

After completing her Ph.D. on the mating behavior of lobsters in Jelle Atema's lab at the Marine Biological Laboratory in Woods Hole, Diane Cowan had become a professor of marine science at Bates College in Lewiston, Maine. Diane liked to escape to the coast from landlocked Lewiston whenever she could. One day she'd been searching for a cove from which to launch her sea kayak when she'd come across a couple of boys playing among the rocks at low tide.

"Whatcha got there?" Diane had asked, peering into the mud.

"Lobsters."

"Huh?"

"Baby lobsters."

Decades earlier, scientists had noted that young lobsters were briefly exposed on rocky beaches during astronomical low tides, when the alignment of the moon with the sun peeled the ocean away an extra few feet. That knowledge had since been neglected because the lobsters weren't easy to see—they hid in the wet crevices between stones. But the boys Diane met that day had been keen observers, and they knew what they were looking at. One of them was the son of a lobsterman.

At first the discovery gave Diane something to show her

class on field trips. She would unveil the lobster babies under the stones, give a lecture on habitat, then drive her students down the road to Cook's Lobster House and order up a round of boiled adults. Over dinner she'd teach her students a lesson in lobster anatomy. For her own dining pleasure Diane would always ask the waiter for a male—more claw meat.

But Diane soon realized that the cove full of babies afforded a rare research opportunity. She'd read Rick Wahle's report on cobble nursery habitats. The baby lobsters she had found were in an inlet called Lowell's Cove, near the tip of Orrs Island, a talon of land jutting ten miles into Casco Bay. It was a few peninsulas to the west of where Rick had conducted his study, but from the look of Lowell's Cove—shallow water, lots of cobblestones, exposed to the sea—Diane thought it must be a lobster nursery. Just as Rick had, Diane could set a square quadrat on the bottom and take a census of the baby lobsters hiding among the rocks. The difference was that Diane could do it all without going underwater.

Diane had earned her scuba certification at the age of sixteen and was no stranger to diving. But a pair of rubber boots made for cheaper and quicker data collection than boats, tanks of air, dry suits, and underwater vacuum cleaners.

In practice, Diane's data-collection scheme sometimes turned out to be more challenging than she had envisioned. That was the case one October afternoon on an excursion to Lowell's Cove, typical but for the fact that Diane was running late. By the time she arrived the sun had set, a bitter wind was whipping up a chop across the water, and the miner's headlamp strapped to her cap projected only a narrow beam of light through the gloom. With a heavy basket of equipment strapped to her back, she picked her way through seaweed-covered rocks to the water's edge.

Twenty minutes later Diane was squatting in the mud, surrounded by cottage cheese containers holding the tiny lobsters she'd collected. In her right hand she held an adapted medical syringe. On the tip of her left thumb was a magnetic chip one millimeter long, engraved with a microscopic bar code. The trick was to get the chip inside the end of the hollow needle without the wind blowing it away. Once she'd loaded the chip, the next task was to implant it inside the leg of an inch-long lobster. Diane aimed her miner's light into the containers at her feet.

"C'mere, lobbie," she sang, plucking a baby lobster from its plastic dish and nearly going cross-eyed with the effort of focusing on its front walking leg—the diameter of a toothpick. Diane had captured fifteen babies by flipping over stones, using the same technique as the boys she'd met on the beach. To minimize disturbance to the lobsters, she planned to return each one to the rock where she'd found it. But the tide had already turned. In a few minutes those stones would be back underwater for another month.

As much as the movement of the tide appears to be a matter of the sea rushing in and out like a frantic commuter, the ocean is, in fact, simply hanging out in the pull of the moon's leisurely, 27.3-day orbit. It is the earth's relentless spinning that accounts for the schedule of the tides. Water near the moon is pulled toward it and water on the far side of the earth falls away, causing a symmetrical pair of bulges in sea level and a corresponding slimming of the sea on the sides. Through this gravitational field the earth churns, dragging its continents with it.

The sun, although very far away, pulls on the ocean as well. When the moon is either at its closest point to the sun or farthest from it—a new moon or a full moon, respectively—the combined gravitational force tugs the bulge of the ocean a little farther from the earth's core, drawing extra water away from the slimming sea on the sides.

For Diane that meant some lifestyle adjustments. Astronomical low tides in Lowell's Cove, Maine, occurred for several days at a time, fourteen times a year, when the sun and moon were located in one of four possible combinations over the longitudes of Cairo, Egypt, or Anchorage, Alaska, or both. Diane usually ended up crouching in the seaweed at dawn or dusk, doing half her work by flashlight. Rain, sleet, or snow, breeze or gale, the tides wouldn't wait.

Bates College indulged Diane's peculiar form of research,

even if she did occasionally show up for class wet and tired after rising at 4:00 A.M. to dig in the mud. A redesign of the curriculum, however, cut the course Diane taught, so she was forced to look for other work. She tried other teaching gigs, but none allowed her to visit Lowell's Cove on every astronomical low tide. So after spending seven years earning a Ph.D., Diane walked into Cook's Lobster House and asked for a job as a waitress. The waiter who'd humored her requests for male lobsters in the past recognized her, and she was hired. The first thing she did was hand the manager a list of the dates and times she wouldn't be able to work during the next twelve months, copied from her tide calendar.

Trying not to rush, Diane rearranged her grip on the tiny lobster she was holding.

"Hold still, sweetie," she whispered. She inserted the hypodermic needle between the pincers at the end of the lobster's leg and pushed until the chip was lodged in the muscle of the animal's forearm. The baby hardly flinched.

The magnetic chip would serve as a tag in case she recaptured the lobster. Implanting it in the muscle tissue was the only way to be sure the tag wouldn't be lost when the baby shed its shell. Whenever she came to Lowell's Cove, Diane brought along her magneto-field detector, a blue box with switches that beeped when it located a lobster carrying a chip. With a pair of scissors Diane would snip off the leg with the chip—the limb would regenerate—and implant a new chip in another leg. Under a microscope, the bar code would reveal when and where Diane had first captured the animal. The tagging data would help Diane determine the rate at which baby lobsters grew and how long they remained in the nursery.

"You're a cute one," Diane said, giving the tagged baby a pat before returning the lobster to its plastic dish. "I hope you show up on my dinner plate someday."

"*Bottom Dollar*, you on there, Jack?"

Jack Merrill slowed the hydraulic hauler aboard his lobster boat and frowned. He recognized the voice, but couldn't place it.

"Yeah," Jack said into his mike, "this is the *Bottom Dollar*, go ahead."

"Jack, it's Bob Steneck. How are you?"

"Hey, Bob! You in the neighborhood?"

By the strength of Bob's radio signal, Jack guessed the scientist was aboard a boat somewhere within fifteen miles of where Jack was fishing off Little Cranberry Island.

"Yup, we're doing some diving today around your island."

"Great. Stop by when you're done, we'll grab a beer and some dinner."

"Sounds excellent. See you then."

Ever since Bob had made his first presentation to the Maine Lobstermen's Association, he'd been impressed by Jack's friendly demeanor. Now, whenever Bob was working off Mount Desert, he made a point of calling Jack on the radio. Those calls served another purpose. Other fishermen listened in, so the conversation became Bob's permission slip to do research in their territory. By now he was wise enough not to steer his boat into a cove dense with lobster buoys without checking with the local fishermen first.

A few weeks earlier Bob had perused a pile of coastal charts with his new intern, a young fellow named Carl Wilson. Carl was an earnest student with a warm smile and a six-foot-four frame. Quick with a joke, he'd grown up in Maine and had spent his summers on Isle au Haut, an island lobstering community twenty miles west of Mount Desert Island. He was familiar with fishermen and eager to get himself wet, whether it was piloting a research skiff through a stiff chop or sucking lobsters off the seafloor with an underwater vacuum. Bob couldn't have found a better ally for the work that lay ahead.

Bob and Carl had studied the charts for coves that might have the same characteristics as Rick's most productive nursery, the western side of Damariscove Island. This ideal site protruded off the coast like a catcher's mitt, facing straight into the southwesterly summer breeze that delivered superlobsters.

Scouring a chart of the waters off Mount Desert, Bob had noticed that the south beach of Little Cranberry was connected by a half-submerged bar to Baker Island in the southeast, creating a mile-long semicircle—it too looked like a sort of catcher's mitt, facing into the summer winds. Bob recalled the hot August day when he'd sat on that same beach with his wife and parents. The beach had been a vast sweep of cobblestones. Maybe on the *Argo* cruise, he and Rick had dived in the wrong places. Bob put Little Cranberry's cobblestone cove on the list.

The more they dove along the coast, the more locations Bob and Carl added to their survey. Two hundred miles to the south, Massachusetts biologists began taking suction censuses, using the techniques that Rick had pioneered. Along with Rick's original nurseries in the western part of the Maine coast, plus the ones he'd added in Rhode Island in the early 1990s, ecologists were soon gathering annual data at more than sixty locations along the coast of New England. Canadian scientists even added a few sites in New Brunswick. The lobster-settlement index the ecologists had envisioned was becoming a reality.

In addition, assistance came from everyday residents of coastal communities. In 1996 Diane Cowan had founded a nonprofit organization called the Lobster Conservancy. She'd quit her waitressing gig at Cook's Lobster House and was now training volunteers to count baby lobsters at low tide. Collaborating with Bob, Rick, and Carl, Diane set up coastal sites to match some of their underwater dive locations and added sites of her own. Helping to coordinate much of this work was an organization called the Island Institute, based in Rockland, Maine. Of the three hundred Maine islands that had been inhabited year-round in the glory days of the cod fishery, most had been abandoned. For the sake of Maine's coastal villages, including the fourteen island communities that remained, scientists, fishermen, and local residents would have to cooperate to make the lobster fishery last.

The level of data collection was unprecedented. By observing patterns in nature and mapping them, Rick, Lew, Bob, and their colleagues were learning to track the health of the lobster population, and soon they might even be able to forecast future trends. They still didn't know exactly why the demographic bottleneck had widened, causing an increase in catches, but they were sanguine that answers could be found.

Before long, however, the ecologists were wrapping their minds around a more vexing puzzle. Catches had continued to rise, but beginning in 1995 Lew noticed a dramatic drop in the abundance of tiny superlobsters that were caught swimming off the coast. At the same time, Rick saw the number of new babies in the nurseries plummet. Off Little Cranberry Island, Bob had finally found a few baby lobsters, but their numbers weren't anywhere near those in the west.

As they patrolled the stony seafloor along the coast, the ecologists felt like hall monitors in a popular school that had suddenly lost its pupils. They kept their observations to themselves until they could gather more data. But if the slump lasted, the consequences for the lobstermen of Maine could be devastating. Had the government scientists been right all along?

PART FIVE

Sensing

13

See No Evil

The eye of the lobster is of such novel and ingenious design that it inspires religious faith and scientific admiration alike. The eyes of most creatures on the planet use lenses to refract light. A lens forces rays of light to pass, at an angle, through a medium that slows them down, thus altering their trajectory. But the lobster's eye, in a design shared with only shrimps and prawns, focuses light by an entirely different principle—not refraction, but reflection. There are no lenses under a lobster's cornea but instead a grid of mirrored boxes. Each box is a long square duct, open at the top and tapering to a point at a package of retinal cells. The four interior walls of the duct are coated with a crystalline lining, and as rays of light enter the open end of the box, they glance off one of the sloping walls at a shallow angle. Like speeding cars grazing a highway guardrail on a gentle curve, the rays of light change direction just enough for them to converge onto the retinal cells at the far end. Each of the lobster's two eyes consists of some thirteen thousand of these tapered boxes.

Religious creationists cite the unique and ingenious construction of the lobster eye as evidence of an intelligent Designer; the surface of the square openings is as geometrically exact as graph paper, and the grazing angles inside each box must be mathematically perfect for the retinal cells to receive light. Scientists agree that this construction is ingenious, and they are using it as the blueprint for a new X-ray-vision space telescope called Lobster-ISS, so named because it will be mounted on the International Space Station. The mirrors that usually focus light in telescopes are useless for focusing X-rays because mirrors simply absorb short wavelengths—with one exception. At a very shallow grazing angle, a mirror can alter the trajectory of an X-ray signal just enough to redirect it without absorbing it. Employing the design of a lobster's eye, the Lobster-ISS will use millions of tiny tapered mirror boxes, fifty to a square millimeter, to collect X-ray images of great swaths of the sky.

Whether a lobster's eye is actually of much use to the animal underwater is another matter. It was this question that Jelle Atema and his graduate students in Woods Hole had in mind when they blindfolded two male lobsters and dropped them into a tank to fight without the benefit of sight.

Jelle's initial research on mating behavior, conducted with Diane Cowan and other students, had raised a host of questions about how lobsters sensed their underwater world. For instance, how did they choose mates, identify competitors, and navigate the neighborhood? Jelle's original goal in Woods Hole had been to study chemical communication in the sea, and he'd chosen the lobster partly for its reliance on its sense of smell. That reliance had been all too clear when Diane had cut off the "noses" of her lobsters—by snipping off their antennules—and watched their mating behavior go haywire. But exactly what sort of information lobsters gleaned by smell, and just how that information was conveyed, were still open questions. So was the degree to which lobsters might be sensing the world through sight and feel.

Experimenting with lobster combat, Jelle and his students had discovered, generated some interesting answers. Jelle had arrived at the study of lobster combat in the lab through a circuitous path that had begun in the sea. Jelle was a devotee of an Estonian zoologist named Jakob von Uexküll, who promoted the idea that animal behavior could be understood only from the animal's perspective. Each organism had its own subjective worldview, which von Uexküll called the organism's *Umwelt*—a sensory environment where the perception of objects and of other animals was defined by the organism's specific needs. A lobster's behavior wouldn't be intelligible to a human observer unless the human knew how a lobster "saw" the world, and not just with its eyes. Jelle couldn't transform himself into a lobster, but in his quest to enter the American lobster's *Umwelt,* he came close to experiencing life as a nocturnal aquatic animal.

Several nights a week over a period of nearly three years, Jelle had donned his wet suit and snorkel and motored a small boat from Woods Hole harbor to a cove on a nearby island. He was usually accompanied by two research associates in wet suits and a support team for surface logistics. As darkness fell the divers slipped into the black water like navy SEALs. Their weapons were low-power flashlights, rubber bands, pencils, and slabs of sanded Plexiglas, which would serve as underwater notepads. Floating facedown on the surface, their lights casting a dim glow over the rocks and eelgrass on the bottom, they watched the lobsters emerge from hiding.

The scientists had drawn up rules of engagement that dictated minimal contact. When the divers saw a new lobster in the cove they would dive, capture it, record its vital statistics, and tag it with a numbered band. Then they let it go and never touched it again. Over nineteen months they tagged 334 lobsters. Three hundred of those lobsters the divers never saw a second time. But about 30 were regular residents of the cove.

Each night the lobsters tracked down and killed prey, including small fish, crabs, snails, clams, and worms. One lobster was caught gnawing on a pork bone. But the animals spent most of their time snooping around the neighborhood interacting with other lobsters. The encounters were less violent than what Jelle and Diane had witnessed in the lab tanks, where elbow room was scarce. But the struggles for dominance were just as intense. Lobsters made frequent visits to the shelter entrances of other lobsters, and larger lobsters often evicted smaller ones. Jelle and his team also saw lobsters wandering among the rocks and through the eelgrass in what appeared to be aimless exploration. But when the scientists tried to catch these animals for tagging, it was evident that the lobsters' investigations had been purposeful. A pursued lobster knew instantly where the closest hiding place was and made a beeline for it. If a second diver moved to intercept it, the lobster had an alternative nook already in mind and immediately changed direction. Similarly, the resident lobsters seemed to know where the occupied shelters were in the neighborhood and who was living in each. On one occasion, the scientists watched a large male emerge from his burrow, run straight to another burrow fifteen feet away, and evict the occupant, a competitor. Shortly afterward the evicted lobster left the cove.

Normally all this would be occurring without even the dim glow cast by the scientists' flashlights. Moonlight and even starlight provided some natural illumination, so the ingenious light collectors of the lobster's eye surely conferred occasional advantage. But could they provide the animals with enough information to identify other lobsters, and the local terrain, over long distances? Probably not. On many a cloudy night in the murky Atlantic, Jelle thought the lobsters he was watching could just as well have been blindfolded without impairing their ability to see.

~

Each of the Fernald lobstermen on Little Cranberry Island possessed an uncanny nose for sniffing out lobsters, but Dan Fernald had the eye of an artist as well. He stepped back from the easel and studied the pastels on his canvas. His painting was a typical island scene—a lobsterman and his two boys hand-lining mackerel off the co-op wharf, the hills of Mount Desert in the background. Nine years had passed since Katy had talked Dan into attending a painting class in 1985, and he'd since developed a style of his own. Dan saw the boreal coast of Maine in colors and shapes that evoked the vibrancy of the tropics. His pictures of the lobster boats in the harbor or the island's clapboard houses by the sea blazed with sunny yellows, flowery pinks, and kelly greens. On his easel now, the wharf and mackerel fishers were bathed in rich red, ocher, and lime. Behind them a sea of turquoise and purple swirls sucked lavender mountains down the sides of the picture.

Little Cranberry's summer scenery had long attracted artists to its shores. One day an artist who'd made the acquaintance of Dan and Katy had asked if he could store a finished painting in the attic of their little barn, which had once housed the fishermen's lobster-trap factory. Katy had mentioned the painting to another visitor and, without really trying, arranged a sale. Before long, Dan and Katy had cleared out the attic and transformed it into a gallery. Some of the Little Cranberry lobstermen thought this was frivolous, even a betrayal of the Fernald family's fishing legacy. But the artists who frequented the island encouraged Dan. He began to display his own paintings alongside theirs, and soon his work was selling for thousands of dollars. When the attic was no longer sufficient, Dan and Katy refinished the first floor of the barn as well. Where equipment for shaping and cutting wire had once served the goal of trapping lobsters, framing tables and glass cutters now served to encase the creations of island artists for display.

Beginning in 1994, Dan could afford to spend more time painting and less time fishing because suddenly he was catching more lobsters. Dan and his fellow lobstermen on Little Cranberry had been hearing about the rise in catches along the western half of the Maine coast for a few years, but now catches had risen in some regions Down East as well. No one knew why.

To celebrate their good fortune the lobstermen converged on the restaurant wharf for a party. Electric guitars, amplifiers, and a drum set were boated out to Little Cranberry Island and hefted onto the dock by a group of musicians. The restaurant proprietor printed up tie-dyed T-shirts with the words "Rock the Dock" and handed them out to the band and the lobstermen. Before the party got under way, Bruce Fernald felt flush enough to host a lobster-eating contest for the musicians. The electric bass player ate thirteen.

After dinner the band began to play. Late into the night, the lobstermen danced with their wives to Rolling Stones songs while teenage boys and girls from summer families eyed each other across the dance floor—a few disappeared to neck on the beach. One woman wore a shirt painted with a pair of lobster claws cupping her breasts. After a few beers, Jack Merrill jumped behind a microphone with his harmonica.

Getting up to go lobstering at five o'clock was going to be murder. In the summertime it wasn't unusual for lobstermen in Maine to wake to a coast enshrouded in fog. If a fog bank blew in before morning, Jack would have an excuse to stay home and sleep. Gathering the dancers into in a ring while the band dropped down to a rhythmic backbeat, Jack gave birth to something like a lobsterman's rain dance.

"Fog!" Jack chanted, encouraging the others as they circled in a kind of drunken Macarena. "Fog! Fog! Fog!"

~

In Woods Hole, the boxing ring in Jelle Atema's lab was a sixty-gallon glass aquarium with an opaque divider down the middle. The combatants, all of them male and equally matched for size, had been purchased from local lobstermen, and prior to fighting they were isolated in separate tanks for forty-eight hours. At fight time the researchers lowered one lobster into each side of the boxing tank and then raised the divider. Without blindfolds or other encumbrances, the two lobsters went at each other with the usual show of strength—aggressive posturing, antenna whipping, body shoving, and even the locking of claws.

If one lobster didn't win within twenty minutes, the pair was disqualified. One lobster usually capitulated before the time limit, signaling its surrender with a display of groveling. But it turned out that the loser's subservience lasted longer than the end of the fight. Jelle and his students staged rematches between the same lobsters twenty-four hours later. As soon as the divider went up, the lobster that had previously lost folded back its antenna, lowered its claws, and cowered. The winning lobster, its superiority already established, left the subordinate alone.

Somehow a lobster was capable of recognizing a former opponent who'd bested him in battle. Attaching blindfolds made little difference. The lobsters' eyes may have been sensitive light detectors, but as Jelle had suspected, they weren't necessary for the animals to identify one another. Jelle and Diane Cowan had shown that lobster behavior in the boudoir was governed by scent. Perhaps in the boxing ring as well, lobsters detected social cues by smell.

On each of four consecutive nights the scientists staged fights between fourteen pairs of lobsters. The fight on the first night established a clear winner in each pair. On the following nights the scientists desensitized the antennules of seven of the pairs by exposing them to distilled water just before the rematch. Sure enough, without their sense of smell the losers failed to recognize their former opponents and unwisely chose to fight a second, third, and even a fourth round, with the same humiliating results.

By contrast, the seven losers who retained their sense of smell recognized the previous winners within seconds and retreated to the corner of the ring, forfeiting every follow-up match. Subsequent experiments revealed that losing lobsters remembered the opponent that had beaten them for as long as a week. It wasn't simply that the losers had become cowards. Nor was it simply that the winners were broadcasting aggressive intent. When a loser was paired with an unfamiliar opponent, even if the new opponent was a dominant male, the loser would still fight ferociously until it lost the first fight—only then would the loser submit. Clearly, the lobsters were identifying and remembering each other as individuals.

Three forms of water current are controlled by the lobster around its body, and these currents are crucial to how a lobster senses the world. At the base of the lobster's walking legs, hidden under the bottom edge of the carapace, are twenty pairs of feathery gills and a series of leaflike fans that draw water into the gill chambers. This powerful current is expelled straight forward from either side of the lobster's head, creating an expanding plume of water that reaches seven body lengths in front of the animal. But the lobster can also hold a pair of its mouthparts just in front of the outflow ducts, deflecting the current backward. With another set of mouthparts the lobster can then fan its own face, generating a current that sucks water from in front of the animal toward its antennules. Finally, the lobster can pulse the swimmerets along the underside of its tail, creating a rearward thrust that can help propel it up a rock face or eject water out the back door of its shelter, drawing fresh water in.

Given the lobster's ability to project a plume of water forward, or to suck another lobster's plume toward its own nose, Jelle suspected that a peculiar quirk he'd noticed in the lobster's anatomy was more than an evolutionary mishap. The American lobster urinates not from some posterior region of its body, but directly out the front of its face. Two bladders inside the head hold copious amounts of urine, which the lobster squirts through a pair of muscular nozzles below its antennae. These powerful streams mix with the gill outflow and are carried some five feet ahead of the lobster in its plume. Quite possibly, lobsters were sensing each other and sending signals—"I beat you up last night, remember?" or "Would you like to mate with me, I'm about to get undressed?"—by pissing in each other's faces.

Finding out, of course, would require a lobster catheter. One of Jelle's students cemented latex tubes to the urine nozzles on a lobster's face with superglue. The two tubes curved over the lobster's shoulders to a Y-connector fastened to the lobster's back. From there a tube ran up to a vial that floated on the surface. Returned to his aquarium and readied for a social encounter with another male, the catheterized lobster wandered the tank, pursued by the floating jar overhead that would collect his urine. The experiment would be akin to sending an angry man into a street-corner altercation with a soundproof hose strapped to his mouth, delivering all his words into a balloon floating above his head.

What the researchers discovered during the ensuing fights was that dueling lobsters accompanied their most punishing blows during combat by intense squirts of piss at the opponent's face. What was more, in scenes akin to a showdown at the OK Corral, the winner of the physical combat almost always turned out to be the lobster that had urinated first. And well after the fight was over, the winner kept pissing. By contrast, the loser shut off his urine valves immediately. In an underwater world navigated by scent instead of sight, the tactic was clearly an attempt to hide.

But what, exactly, were lobsters doing when they peed at each other during fights? One possibility was that the lobsters weren't communicating at all but instead using urine as a kind of chemical weapon—some insects inject caustic chemicals into their urine and pee at predators as a form of defense. But that seemed unlikely. There were two other possibilities.

Dissections in Jelle's lab had revealed an unexplained gland inside the lobster's head that emptied directly into the urinary tract. Conceivably, the gland was used to produce proteins that registered a unique odor signature for each individual. Indeed, a urine analysis of several different lobsters revealed a unique pattern of molecular weights in the protein content of each animal's pee. Mice are known to employ proteins in urine to convey individual identity, but a similar system has never been shown to exist in invertebrates. Perhaps what lobsters were doing when they squirted urine during fights was simply the equivalent of the angry man on the street corner shouting his own name every time he punched his opponent. "This pain goes with this name," the lobsters seemed to be saying, "so remember it well."

The second possibility, even less certain, was that victorious lobsters might be reinforcing their own aggression through increased reception or production of a chemical motivator such as serotonin. The result might be to instill a sense of confidence, so to speak. Chemically, this condition wouldn't be unlike the enhancement a human gets from a cup of coffee—caffeine delays the reuptake of serotonin, which makes more of the motivating chemical available in the brain. A dominant lobster with enhanced chemical motivation might exude some sort of chemical clue to the heightened level of aggression in his urine, which in turn could be detected by the animal's peers and paramours.

Jelle and one of his students designed an experiment to test whether other lobsters could detect something like a sense of confidence exuding from a winning male. First they forced a pair of males of equal size to compete for a single shelter in a tank. As expected, one of them emerged as the dominant animal and the shelter's sole occupant. Meanwhile, at the head of another tank called a flume, the scientists installed two shelters behind a sheet of wire mesh. Jelle's flume was nine feet long, and it had a water current flowing straight through it from one end to the other. The scientists then separated the two shelters with a wall that ran halfway down the flume, creating a pair of parallel waterways that emptied into a single space downstream. One male was then ensconced behind bars in each waterway upstream, at the top of the flume. From a pool of twenty female lobsters, none of whom had encountered either of the males before, the scientists placed one female at a time in the tank downstream, where the waterways converged. The female thus had two streams of scent to choose from, one originating from each of the unfamiliar males.

Initial trials, with the males catheterized, revealed that when a female drew near one of the shelters, the male inside it pissed with vigor out the door at her. When the catheters were removed and the urine could flow freely down the flume, the females spent most of their time trying to push through the bars of the shelter containing the male that had proved himself dominant in the other tank. Since the females hadn't met him before, the only way they could have known of his superior status was by detecting some indication of the sense of confidence he exuded. Presumably, his piss smelled better.

On subsequent trials the scientists compared the attractiveness of two males that were both dominants in their own right in separate tanks. One of them, however, had won a greater number of fights than the other. The females in the flume had no way of knowing that, and yet they consistently preferred the male that had won the greater number of fights.

Jelle remembered the first time he had imagined how lobster mating worked, the female poised to receive suitors in her shelter, emitting a sexy scent and attracting males from downstream. For nearly twenty years Jelle had followed the trail of chemical communication in the sea. He still didn't know exactly what combination of individual identity and general confidence gave a particular lobster its social status. But it was clear that in a natural setting, the basic pattern of lobster mating was just the opposite of what he had first envisioned.

The dominant male waited in his shelter, peeing out the door of his apartment at the females who came calling. A female would poke her head in and pee back at her prospective mate, a love potion in her urine suppressing his bellicosity and putting him in the mood for courtship. He would stand on tiptoe and pulse his swimmerets, drawing her urine in and fanning it appreciatively about the boudoir.

The scents of his dominant masculinity and her seductive femininity would mingle, Jelle supposed, and waft out the back door of his shelter, like an olfactory billboard posted in his backyard. With this billowing advertisement to the females of the neighborhood that his love nest was active, it hardly seemed surprising that a dominant male might develop a sense of confidence.

14

Against the Wind

For all of Jelle Atema's efforts, he had yet to truly enter the lobster's *Umwelt*—to sense the world as a lobster did. The problem was that chemical signals were a messy business. Eyes detect light that mostly travels in straight lines at constant velocity. Ears detect acoustic waves that disperse through a medium according to well-established rules. But smell requires the detection of patches of molecules carried by ever-changing eddies inside chaotic plumes of moving air or water.

The first time Jelle had ever dropped a chunk of fish into a lobster's tank, he'd been amazed by the speed with which lobsters could pinpoint an odor source. At the first sign of an attractive scent they flicked their antennules and began to move, walking slowly at first and adjusting their heading with an accuracy that was nearly instantaneous. After a few seconds they were jogging toward the origin of the odor, whether it was a tasty morsel of food or an alluring lobster of the opposite sex. How did they do it?

The lobster's nose—its pair of antennules—is perhaps a less remarkable organ than its eyes, but the antennules are far more useful underwater. Attached to them are hundreds of sensory hairs with permeable walls. Inside each hair is a dense tuft of some four hundred grasslike neural cells that are attuned to particular combinations of molecules. If a lobster is walking into a current, the animal aims the antennules straight up like rabbit ears, so that water washes against the full length of the sensory-hair array. But if clear reception is interrupted by confused currents or imperfect orientation, the lobster flicks its antennules downward in swift strokes to obtain a stronger signal. Thus the analogy to sniffing.

Detecting waterborne chemicals is one thing; tracking them to their source is quite another. The question of how lobsters use their antennules to locate the origin of seemingly random splotches of whirling odor had dogged Jelle for decades. He'd spent hours crouched over a drainage gutter in the floor of his basement lab, squirting dye into the flow of seawater in a vain attempt to detect patterns in the plume of swirls. It was an impossible task because he couldn't see the plume from a lobster's point of view.

That changed when Jelle learned of a research project under way in Colorado. A neuroscientist in Denver was working on ways to repair the neurotransmitters of patients with Parkinson's disease. He had developed a miniature electrode for detecting the neurotransmitter dopamine in rat brains. The electrodes were about the same size as the receptors on a lobster's antennules. Jelle guessed that a pair of these detectors might be able to "see" a water current containing dopamine the same way a lobster "saw" a current containing an odor.

Jelle began to collaborate with the neuroscientist and ran an initial trial in his flume tank. He injected a jet of dopamine into the current from a nozzle at the head of the tank. Then he positioned one of the electrodes at sixty successive locations across a grid downstream, recording the fluctuations in dopamine concentration at each location. As water flowed through the tank, it dispersed the dopamine in an increasingly turbulent plume of cascading eddies. But this time Jelle wasn't watching the plume with human eyes. Instead, he was seeing what a lobster standing at different locations in the tank would sense with its antennules—a series of changes in chemical concentration.

Neurological studies in Jelle's lab had determined the detection frequency, response speed, and acclimation rates of the chemoreceptor cells in a lobster's antennules. Using this data, Jelle calibrated the recordings from the dopamine detector to match the lobster's own sensory abilities and graphed the results. The effect was stunning. The odor landscape inside a turbulent plume looked to a lobster something like a mountain range of chemical peaks, each following the next in time. Near the source, those peaks would be tall and steep because the patches of chemicals passing across the lobster's antennules were dense and had clearly defined edges. Farther away from the source, or off to the side, the peaks the lobster would perceive were by degrees shorter and gentler in slope, and spaced farther apart, because the patches of chemicals had become fuzzier and more diffuse.

Dye swirling in a current might appear chaotic to the human eye, but after several hundred million years of tracking odors underwater, a lobster inside a turbulent odor plume surely felt right at home. Most of Jelle's colleagues thought the idea was crazy, but Jelle believed a lobster might be capable of pinpointing its own location in relation to the distant source of the scent simply by the look of the chemical slopes and hills in its immediate vicinity.

The fact that lobsters had not one but two antennules, spaced a body width apart, was also essential to tracking odors. That was all too clear when Jelle snipped one antennule off a lobster and it started walking in circles. One of Jelle's students set about constructing a lobster backpack that contained a submersible amplifier and two dopamine electrodes. When the backpack was attached to a lobster, one electrode sat directly behind each antennule. The lobster was blindfolded and lowered into the downstream end of the flume tank, and a brew of dopamine and squid extract was squirted from a jet upstream. A cable from the lobster's backpack supplied a computer with the electrode readings while another cable supplied the computer with a video feed of the lobster's movements, filmed by an overhead camera. As the lobster tracked the scent of squid up the tank, the computer synchronized the chemical and visual data. The scientists could see the pulses of odor the lobster was experiencing on the right and left sides of its head while it was deciding which direction to turn. As expected, the lobster turned toward the antennule that detected steeper and higher hills of odor before the other antennule did, enabling the lobster to "climb" the mountain range of the plume to the source, the animal's trajectory growing ever more accurate as the steepness of the peaks increased.

Perhaps it was inevitable that the next experiment would involve mounting two odor-release nozzles on the lobster itself, one pointing at each antennule. Now that Jelle understood how a lobster perceived an odor landscape, it was a simple matter to generate lobster virtual reality. A hungry lobster could be steered through an empty tank at will by squirting bursts of fish-flavored seawater at one antennule or the other.

And after lobster virtual reality, perhaps it was inevitable that lobster artificial intelligence would be the next hurdle. When the diminutive automaton was complete, poised at the downstream end of Jelle's flume tank, it had 256K of RAM and its name was RoboLobster.

~

The hurricane warning crackled over Bruce Fernald's radio aboard the *Double Trouble* two days before the end of August 1996. The storm was a monster and approaching Little Cranberry Island quickly. Of Bruce's eight hundred traps, a third were still sitting in shallow water near shore, where a gale could beat them into tangles of wire and twine.

Already, one of those traps had nearly killed him. Bruce had been breaking in a new sternman aboard the *Double Trouble* that year—a young man who'd never worked on a boat before. Two weeks before Bruce's forty-fifth birthday, the fellow had lost his grip on a brick-laden trap at low tide and let it fall fifteen feet from the wharf to the boat. The trap had missed hitting Bruce by a yard. Bruce was glad to have escaped injury, not least because catches were still on the rise and he couldn't afford to sit out the height of the trapping season. But with a hurricane on the make, Bruce would have to remove his gear from shallow water to ride out the storm.

The next two days passed in a blur of coiled rope and

stacked traps. Aboard the *Double Trouble* Bruce and his sternman, like worshipers of some Pharaoh of lobsters, built pyramid after pyramid of wire-mesh rectangles as they hauled the dripping traps from the water and lugged them in boatloads to land. Little Cranberry's fleet of old pickup trucks came and went from the wharf in a chorus of squeaky springs as the fishermen carted their traps up the road.

The day the hurricane was to arrive a stiff breeze raced across the harbor. The men spent the afternoon hauling small boats out of the water and battening down equipment on the co-op wharf—electric scales, hundreds of wooden lobster crates, dragnets, and plastic bait trays. After checking the mooring lines to the lobster pens, the lobstermen rowed their skiffs into the harbor and double-checked the mooring lines to their lobster boats, which were too big to pull onto land on short notice.

When the rain came crashing down in leaden sheets across the harbor and the ocean frothed white out of the west, the fishermen knew they were inside the leading edge of the hurricane, and there was nothing more they could do.

Still in their rain slickers and dripping wet, they congregated in the bar at the end of the restaurant wharf to watch the storm come. It was the final day of the restaurant's summer season, when the owners held their customary closing night for the islanders—no tourists allowed. Leftover beer would flow for free until the kegs ran dry. Clutching pints of Harpoon ale and Budweiser, the fishermen sat with their backs to the bar, gazing out through the windows while the rising tempest buffeted the wharf on its pilings and pulled their pitching boats tight on their mooring lines.

One of their fellow fishermen had been running last-minute errands on the mainland, and they wondered whether he would attempt to return to the island. Not long afterward, through the gray dusk they noticed his boat approaching from the north, plumes of spray flying. The fishermen stood up from their barstools and strode to the rattling windows, their pints of beer forgotten, while a part of them, a descendant of the great cod fisherman Sam Hadlock and one-fifteenth of the whole that made up the island fleet, plunged laboriously toward home. He was half a mile away in jagged peaks of surf.

Suddenly the boat turned east and headed out to sea. They knew what the man was doing. He was afraid the boat might roll on its beam and swamp, so he would try to surf the waves downwind on an angle, find a trough where he could gun the engine and spin the boat back to the southwest, and pound upwind into the oncoming walls of water, splitting them with his bow and making a slow, zigzag kind of progress until he found the harbor. It took him half an hour to cover that half mile, but he managed to reach his mooring and put the boat on its tether. When he walked into the bar he was soaked to the skin, but he had a big grin on his face. A cheer went up and a beer made its way into his hand.

~

Little known to the general public, in a nondescript building on the northern fringe of the campus of the Massachusetts Institute of Technology is a small laboratory where research has been funded in part by the U.S. Navy. Inside, scientists have constructed torpedoes that can all but think for themselves. They are called AUVs, or autonomous underwater vehicles, and they disappear into the sea and carry out missions without remote control. Oceanographic research and oil exploration are among the civilian applications of these devices, but close cousins of these machines now assist the U.S. military in amphibious assault operations as well. During Operation Iraqi Freedom several were dropped overboard in the port of Umm Qasr. On dives that could last twelve hours or more, they swam free on their own recognizance, hunting antiship mines.

Thomas Consi, a member of the AUV group of researchers at MIT, was a biologist by training but he liked having little robots walking across his desk. After a day designing intelligent torpedoes Tom would stop by a toy store and hunt for inanimate objects he could bring to life. When MIT hosted its first Artificial-Intelligence Olympics, Tom fielded a toy army tank that he'd taught to follow a beam of light.

The term for this sort of work, and for the projects under way in the AUV lab, is biomimetics—the mimicking of natural physical and behavioral structures using artificial devices and algorithms. The goal isn't only to create useful robots but also to gain insights into the biology of living organisms. Some of Tom's colleagues were constructing a beast called RoboTuna. The purpose of the project was to understand the complex hydrodynamics involved in how a real tuna swims.

When Tom heard about Jelle Atema's plume-tracking project with lobsters, he invited Jelle to MIT, and the concept for RoboLobster was born. Working alongside the intelligent torpedoes in the AUV lab, Tom and a colleague machined, assembled, and programmed the device, and soon the lobster-sized robot had made its way into Jelle's nine-foot flume tank in Woods Hole.

RoboLobster sat poised like a jet fighter on a runway, ready to attack the oncoming current. His watertight body was a shiny cylindrical hull that housed an onboard computer with a 20-megabyte hard drive and sixteen AA batteries. Like an airplane, RoboLobster's fuselage was topped with blinking red lights. A pair of direct-current motors powered his little rubber wheels and provided steering. Protruding straight up from RoboLobster's head were two stainless-steel wires—a pair of metallic antennules that served as conductivity electrodes. Fresh water was running through the flume instead of salt water, and instead of dopamine or fish juice, RoboLobster would be tracking the scent of a salt-and-ethanol mixture injected into the current.

The brain of a real lobster consists of several nerve ganglions strung together, and nearly half of their volume is dedicated to processing the signals collected by the animal's sense of smell. RoboLobster's brain was far simpler, and the only thing he had to process was smell. He was programmed to turn right or left depending on which electrode detected a higher concentration of salt in the water, to travel in a straight line if the concentration was similar on both sides, and to move backward if he lost the scent.

Inside the flume RoboLobster managed to track down the source of the salt about 25 percent of the time. Compared to the swift efficiency of a lobster, the paths RoboLobster traveled were torturous. Still, RoboLobster was a start, and Tom was proud of him. RoboLobster was one of the first biomimetic automatons to function successfully underwater. And what was perhaps most startling was how "biological" RoboLobster's performances looked. When plotted on paper, the robot's tracks reflected more the fluid complexity of nature than the programmatic code inside RoboLobster's microchip head. The implication was clear. A real lobster didn't have to be very smart to find its way around inside the currents in the ocean. It just had to be equipped with the proper detectors.

What the experiments suggested to Jelle was that a robot equipped with additional sensors for detecting the direction and speed of water flow might nearly match the skills of a real lobster. No actual animal tracked odor without reference to the movement of the medium containing that odor, be it air or water. The hairs on many parts of the lobster's body detected touch or motion, so water flow was clearly a constant part of the animal's sensory experience. Jelle guessed that a lobster's antennules were involved in flow detection as well.

Jelle and a student snipped the antennule off a live lobster and clamped the antennule upright inside a sort of wind tunnel constructed from a section of clear plastic pipe. They flooded the pipe with water and, while taking high-speed film of the antennule through a microscope, ran oscillating water currents through the pipe at different frequencies. The scientists discovered that different parts of the antennule itself resonated with different frequencies, like a guitar string of varying pitch. Jelle had been an avid flutist for decades—once he'd even played a tune on an old lobster claw for an audience at a meeting of the American Association for the Advancement of Science. For Jelle, the notion that lobsters might be not just smelling but, in a sense, listening to the symphony of the ocean's currents was entrancing.

Elsewhere the quest for a robotic lobster had taken a more sinister turn. The U.S. Navy was now considering plans for a beachhead assault that would begin with thousands of biomimetic lobsters dropped offshore from low-flying aircraft. Clambering over rocks and sniffing their way through currents toward shore, the lobster robots would search out mines and blow themselves up on command. Soon the Pentagon was funding robotic-lobster research to the tune of several million dollars.

The work was carried out in a lab at the end of a narrow peninsula on the northern coast of Massachusetts, accessible only by a two-lane causeway. There, a bank of computers belonging to Northeastern University's Marine Science Center analyzed video feeds of lobsters walking on treadmills in glass tanks. Once lobster motion was translated into computer code it could be downloaded to microprocessors and fed via electrodes to artificial leg muscles made of nickel-titanium alloy. Fleets of thousands of robotic lobsters scurrying across the seafloor could have civilian applications as well as military uses, such as patrolling for pollution.

And who knew—if the natural lobster population was being overharvested and the fishery had to be shut down, perhaps the federal government could task New England's lobstermen with catching and disarming explosive automatons over which the navy had lost control. It wouldn't be that much more dangerous than what they already did for a living.

PART SIX

Brooding

15

Gathering the Flock

The old ship, a thousand tons of steel seesawing slowly on the undulating sea, groaned with the weight of the trawl. The twin electrohydraulic winches hummed with the strain of six thousand feet of metal cable. Scientists from the National Marine Fisheries Service waited under the yellow crane in the stern, bundled in orange overalls and rubber boots. They weren't sure what the net would haul up. The aft deck of the R/V *Albatross* was a thousand square feet, and the trawl could pile it ankle-deep with spiny redfish. Or it could splatter out the occasional cod, huge flat halibut, or hideous hake with its stomach popping out of its mouth from the change in pressure. The *Albatross* had once caught a streamlined gray object that looked like a shark. It was the fuel tank of an F-16.

Over the stern clanked the green metal doors that held the trawl open underwater, and the net disgorged its catch. A waterfall of silver butterfish laced with quivering pink squid spewed onto the deck. In their midst lay a mammoth lobster, each of its claws alone a foot long. Dropping to their hands and knees, the scientists sorted the butterfish and squid into baskets for counting and measurement. The lobster too would have its size, sex, and location recorded.

Owned by the U.S. National Oceanic and Atmospheric Administration, the R/V *Albatross* spends 250 days a year sailing U.S. waters in the North Atlantic, trawling for ocean life so that federal scientists can generate an ongoing census of commercially valuable species, including the American lobster.

After several weeks offshore the *Albatross* steams back to its home port at the National Marine Fisheries Service science center in Woods Hole.

Catching and counting sea creatures aboard the *Albatross* is grueling work—the shift schedule is six hours on, six hours off, twenty-four hours a day—but it can be a welcome change from the office. That is especially true for the scientists responsible for managing the lobster fishery. When they are away from their desks, off catching lobsters in the Gulf of Maine, they can escape the frustrations of managing an industry that claims there is no downside to hauling in the highest catches in history.

In 1994 Maine's lobstermen trapped thirty-nine million pounds of lobster, nearly double the historical average of twenty million pounds. Alarmed, a new committee of scientists convened and issued an official government assessment of the lobster stock, which warned that lobsters were being overfished. The fact that catches had nearly doubled was not cause for celebration, the scientists felt, but for serious concern. The situation brought to mind the history of the cod fishery, in which an exponential rise in the catch had been followed by a devastating biological and economic collapse.

Bob Steneck was worried too. The decline in superlobsters and baby lobsters that Lew Incze and Rick Wahle had seen in western coastal Maine since 1995 had continued, and nursery settlement remained dismal. Bob and his fellow ecologists were forced to ask themselves the obvious question: Had the decline been caused by a drop in the number of female lobsters making eggs?

For more than a decade, government scientists had been warning that too many female lobsters were being trapped too soon to produce enough eggs to sustain the lobster population. If that was the case, Bob reasoned, the R/V *Albatross* ought to have been catching fewer and fewer mature female lobsters when it dragged its net across the bottom. Bob contacted the National Marine Fisheries Service and asked to see the *Albatross* data.

The *Albatross* data was of particular interest to Bob because counting large lobsters was more complicated than taking a census of baby lobsters in their nurseries. Tagging studies have proven what lobstermen have long known to be true. As lobsters mature they begin to migrate seasonally, often walking twenty miles in a year. Larger lobsters can cover far greater distances. One lobster tagged and released near the Canadian border was later caught off Rhode Island—it had jogged across five hundred miles of mountainous terrain in six months. It was these migrating monsters that Bob was after because they were the ones that produced the most eggs. The realm of the ocean they inhabited was too deep for scuba gear, but the net of the *Albatross* was capable of dragging them up.

Though Bob's tax dollars were helping to pay for the work of the *Albatross*, the reception he received from the National Marine Fisheries Service was chilly. Some of the same scientists who had disagreed with Bob during the fight over the minimum size still worked there, so perhaps it wasn't surprising. Bob pleaded with a top official at the agency, then with a congressional aide, but no data was forthcoming.

At another meeting of the Maine Lobstermen's Association, Bob got to talking with Jack Merrill. Jack had been promoted to the association's vice presidency, and he was as concerned as ever to ensure that the lobster population remained healthy. He thought that trying to obtain the *Albatross* data was a waste of time.

"I doubt the federal trawl survey would tell us much," Jack said. The only places the *Albatross* could tow its net were in flat expanses far from shore, where the net wouldn't tangle on trap lines or rocks. "They're certainly not counting the egg-bearing lobsters that we see in our traps."

Bob agreed that Jack might be right. The number of lobsters the *Albatross* caught in a year was about 150. That wasn't much of a sample size.

Indeed, given the limitations of the trawl survey, it was perhaps little wonder that government scientists took a dim view of the lobster population's capacity to produce eggs. One government scientist had declared that according to his calculations, the number of V-notched lobsters in the ocean couldn't be much more than ten thousand. When Jack had heard that figure, he hadn't known whether to laugh or cry. He cut V-notches in that many lobsters himself every two years.

"It's ridiculous," Jack complained to Bob. "One lobsterman can haul up more V-notched females in his traps in a day than the trawl survey picks up all year. Why aren't scientists counting those?"

~

The lobstermen of Little Cranberry Island had been trying to get scientists aboard their boats ever since Warren Fernald had invited Maine's commissioner of marine resources aboard the *Mother Ann*. With her coffee cans, Katy Fernald had given the Little Cranberry fishermen a way of counting V-notched lobsters themselves. Then the Maine Lobstermen's Association had begun its postcard survey, asking members to scribble down the number of eggers and notchers they caught over a two-day period every autumn. The MLA survey had now been operating for a decade, but even Jack Merrill recognized that the survey had an Achilles' heel. Anyone could argue that the lobstermen were lying.

Meanwhile, in New Hampshire, ecologists had been experimenting with other methods of surveying the lobsters that came up in traps. In the 1980s government scientists in Maine had tested something called "sea sampling." Researchers had boarded lobster boats to record data not only about the catch, but more importantly about the lobsters that were returned to the ocean. The trials were too limited to be effective—they had operated out of only three harbors once a month—but the idea was sound. The New Hampshire scientists were supplementing similar sea-sampling trips with logbooks filled in regularly by fishermen. Checking the lobstermen's data against the sea-sampling data helped ensure accuracy.

Collaborating with colleagues in New Hampshire, Bob Steneck distributed logbooks to fishermen in Maine so they could record the lobsters they hauled up over an entire season. Bob put his tall, blond-haired assistant Carl Wilson in charge of the project. After his first summer as an intern, Carl had become Bob's graduate student, and had developed an easy rapport with Bob's fishermen friends. But when Carl phoned the participating lobstermen to collect data, the limitations of the logbook system were apparent.

"How's your logbook coming along?" Carl would ask.

Often the lobsterman would make pleasant conversation for a few minutes, then apologize that the logbook seemed to have blown overboard. There was an exception. Jack Merrill's logbooks arrived regularly in the mail, numbers penciled in neat columns—date, depth, number of traps hauled, total number of lobsters, number of oversize lobsters, number of egg-bearing lobsters, and number of V-notched lobsters.

To supplement Jack's logbooks, whenever Bob and Carl were diving off Little Cranberry Island, Bob would call Jack on the marine radio as usual, but now they would arrange to meet at sea for a session of sea sampling. Carl would climb aboard the *Bottom Dollar* carrying a couple of plastic fish trays, a pair of calipers, and a notebook. As Jack hauled through his strings of gear he dumped the lobsters from his traps into one of Carl's trays. Carl would hunch over the lobsters one by one, measuring them with calipers and recording sex, V-notch, and egg-bearing status before handing them off to Jack's sternman.

Carl realized that most fishermen were too busy to fill out detailed descriptions of their catch. Maybe what Carl needed wasn't logbooks, but an army of sea samplers. Back on land Carl rounded up a team of summer interns and trained them in the sea-sampling protocol. In the summer of 1997 they started talking their way aboard boats up and down the coast. Outfitted with rubber overalls, heavy work gloves, life vests, and tape recorders instead of notebooks, Carl and his interns cajoled, joked, danced, and sometimes puked their way through long stints of lobster fishing aboard nearly a hundred vessels. Shouting into their recorders over the din of diesel engines, they measured and sexed several thousand animals a day.

The effort was such a success that Carl's sea-sampling program earned the institutional backing of the Island Institute and a hundred-thousand-dollar appropriation from the United States Senate. Things had come a long way since Katy Fernald's coffee cans. Jack was thrilled, and his fellow fishermen were eager to get involved. When sea samplers scheduled their next visit to Little Cranberry Island, Bruce Fernald signed up.

Bruce steamed over to the mainland first thing in the morning. Steering the *Double Trouble* toward the wharf, Bruce decided he could get used to doing science. There was nothing like pulling your boat up to the float at 6:00 A.M. and seeing a couple of college girls in rubber overalls eager to jump aboard.

~

To Jack Merrill there was a problem with the new sea-sampling program, and he pointed it out to Bob Steneck. If Bob's goal was to judge whether the lobster population was producing enough eggs, surveying during the summer probably wouldn't suffice. It was in the autumn that the real magic happened.

Every fall Jack prepared for "the gathering of the flock," as he liked to call it. As the first wave of soft-shelled lobsters began to migrate offshore in September, Jack pulled up his traps, untied the shallow-water ropes, stacked the gear in the stern of his boat, and set a course for deeper water. He tied on longer ropes and reset the traps in thirty or forty fathoms in the muddy valleys where the shedders would move offshore. Then, come early October, he'd see a sudden burst of mammoth females. Their shells were rough and battle-scarred. Most of them were V-notched, and the undersides of their tails were loaded with eggs.

At first, these venerable females would show up in Jack's traps in deep water—250 to 300 feet—almost as if they were coming from far offshore. He'd see a few at a time, and he'd toss them back overboard. Soon he'd see more, and in shallower water—150 or 200 feet—as if they were moving toward land like an invading army, and he'd throw them back again.

Most of them seemed to stop several miles from land and encamp there for two or three weeks. Then they turned around and headed back offshore. That's when the magic began.

For the next week or two, in late October and early November, large V-notched lobsters swarmed into traps in a massive wave heading back out to sea. As the peak of this migration passed through the fishermen's gear, each lobsterman could catch hundreds of big notchers and eggers in a single day. The animals gorged themselves on the bait and attacked anything that moved, including humans. The crusher claw of a five- or six-pound mother lobster wasn't something a fisherman could allow his hand to get stuck in, ever. Gingerly Jack would pry the V-notched mothers from his traps and drop them over the side, knowing he'd probably catch them again a few days later in deeper water.

These female hoards of autumn took a toll on Jack, whether in bushels of expensive herring or hours of hazardous handling. But they were the fishery's future. They also provided opportunities for informal research. From time to time Jack would haul up a big notcher with a message carved onto her shell—usually including a date and the initials of one of his fellow fishermen. Jack would then call his colleague on the radio to determine how far the animal had traveled. The autumn run of eggers even provided the occasional opportunity for entertainment. Once, the monotony of Mark Fernald's day was broken when he hauled up a big notcher that was dressed in a Barbie-doll outfit, complete with high-heeled sandals. Mark had to lift her skirt to see her V-notch. Another Little Cranberry lobsterman caught her again about a week later—she had walked three miles in heels.

Following the veteran females, Jack would usually witness a second surge of younger females. These were the new mothers—shiny two-pounders with unnotched tails and glistening masses of dark green eggs. These lobsters were well over the minimum legal size, and how they had escaped the fishermen's traps until now was still a mystery to Jack. But for a brief week or so, while they followed their elders offshore, the new mothers entered traps with abandon. Jack and his fellow fishermen plucked them out, cut a V-notch in their tail flippers, and set them free.

Around the first week of November, after moving through Jack's strings of traps, the eggers and notchers disappeared into the depths almost as abruptly as they had come. Following behind them, the autumn's second burst of shedders began—these were mostly adolescent lobsters that had just molted up to the minimum size, on their way offshore for the winter.

"I do think it's some sort of gathering of the flock," Jack told Bob Steneck. "Those big females are the wise old ladies of the lobster population. They're not coming inshore to shed their shells or lay their eggs, so maybe they're coming in to lead the young ones out to the wintering grounds. You know, teaching them how to migrate."

"You might be right," Bob sighed. "I have to say, I have profound respect for the stupidity of lobsters. But I've been wrong before."

In response to Jack's prodding, Carl Wilson returned in early November and brought along a videographer to document the catch. The men spent the night at Jack's house. It was pitch-black outside when he woke them. Bundled in waterproof jackets and wool hats, they boarded the *Bottom Dollar* and roared offshore. After nearly an hour of travel Jack arrived at his first string of traps and started hauling triples—sets of three traps on each buoy—in two hundred feet of water.

"For twenty years," Jack yelled over the whine of his hydraulic hauler, "I've been trying to tell scientists what I see out here. The state has no statistics on it because they get their data from dockside landings, not from what we throw back overboard."

Carl nodded. During his sea-sampling trips that summer, he'd been impressed with the number of V-notched and egg-bearing females that lobstermen picked from their traps and returned to the sea. In July and August Carl had frequently counted as many as ten V-notched lobsters a day.

Jack's first trap broke the surface. It had been sitting on

the bottom for five days, but it didn't have much in it, nor did the rest of the string. Jack blamed the full moon and the strong tidal currents it caused, which might have discouraged lobsters from leaving their shelters.

But Jack's next string was in deeper water, and a single trap came up carrying seven large egg-bearing females, most of them with V-notches, plus one V-notched lobster without eggs. Carl started recording data. He didn't stop for the next eight hours. Trap after trap came over the rail loaded with two-pound, three-pound, and four-pound eggers, mostly V-notched. While Carl's counting trays overflowed with angry lobsters waiting to go back overboard, only a few lobsters made it into the holding tank of animals that Jack would sell at the wharf.

"Hey, look," Jack would say when a legal lobster came up in a trap, "one I can keep!"

In one triple there were twenty-five lobsters, only three of which went into the holding tank. Nine of the remainder were notchers or eggers, and the rest were either too small or too big. Later Jack hauled up several monster males, six and seven pounds apiece, all of them well over the maximum-size limit.

"That's a nice bull," Jack would comment, before dropping a bulging male overboard.

One notcher came up with the letters "MF" carved on her back—Mark Fernald. Jack got on the radio and learned where Mark had caught the animal. Another notcher appeared with a numbered band attached to its wrist, put there by a lobsterman farther Down East. Carl jotted down the information printed on the band, which he would use to contact the fisherman later.

Carl reached for a bundle of yellow tags of his own, each printed with an ID code and his telephone number. He laid a row of V-notched lobsters upside down on the deck, their tails laden with glistening eggs, and tied a tag around the wrist of each before dropping them over the side. Lobstermen who caught them later could call Carl, generating data on the lobsters' movements.

Earlier that fall, Jack had hauled up a large egger that Carl's sea-sampling team had tagged back in August, aboard a boat that worked more than forty miles from Little Cranberry. Jack had phoned Carl and the two men had discussed the lobster's location. Based on the typical size and movements of a lobster's legs, they estimated that the animal had walked the equivalent of Maine to Florida for a human in just over a month.

When Carl had finished tagging, he straightened up and leaned against the aft wall of the *Bottom Dollar*'s cabin while Jack ran the boat to his next string. Carl looked dazed.

"This is one of the highest concentrations of brood-stock lobsters I've ever seen," he said. He nodded to himself in amazement. "It's pretty impressive."

Jack slowed the boat, gaffed his next buoy, and wound the rope into the hauler.

"This is a typical day for this time of year," Jack said. "The other guys in the harbor and I did some rough calculations. We estimated that together we must cut V-notches in fifty thousand lobsters a year." That worked out to about four thousand notches cut by each of the Little Cranberry lobstermen every year.

By the end of the day Jack had caught 984 lobsters. Of those, 716 had gone back into the sea.

Carl, instead of counting ten or so V-notched lobsters as he would have during the summer, had counted more than four hundred, nearly half of them carrying eggs. In addition, Jack and Carl had cut V-notches in another 174 new mothers—females that had never been caught before and had just extruded eggs, probably for the first time.

Carl thought the number of eggers he'd seen that day was mind-boggling. What was mind-boggling to Jack was that the government still considered this a population suffering from a lack of eggs.

~

Bob Steneck was a *Star Trek* fan. If the big V-notched females appeared only in the fall, he liked to think it was because the rest of the time they employed the lobster equivalent of a Romulan cloaking device.

Female lobsters carrying eggs undergo a behavioral shift that helps protect their spawn not just from predators but from attacks by other lobsters. Egg-bearing females are more reclusive, but if challenged, they are quicker to dispense with niceties like claw lock and kill their opponents outright. Bob remembered that in his experiments with neighborhoods of pipes, large lobsters had withdrawn when faced with the prospect of defending their territory against an overwhelming number of smaller lobsters. Surely, even a mammoth female would avoid other lobsters in order to protect her brood of seventy thousand or eighty thousand eggs.

A remarkable new piece of evidence, garnered by a colleague of Bob's working at the University of New Hampshire, suggested that the best way for a lobster to avoid antagonistic encounters with other lobsters was simply to avoid fishermen's traps. Winsor H. Watson III, a professor of zoology, came up with the idea of videotaping lobsters around a lobster trap. He had worked on several other projects, including an indoor lobster racetrack—the gambling was restricted to Win's technicians and graduate students—and a lobster treadmill with an Astroturf belt for monitoring heart rates during exercise.

The breakthrough for studying lobster behavior around traps came when Win and some of his students invented what they called the "lobster-trap video," or LTV. It was a regular trap with a kitchen containing bait and a parlor, except that it was also outfitted with a camera that looked down through a Plexiglas roof, a waterproof VCR unit, and a red LED lighting array for night vision. The researchers could set the LTV on the bottom and run it for twenty-four hours to see how many lobsters entered the trap, and what they did once they were inside. The discoveries the scientists made came as quite a surprise to any lobsterman who considered himself a talented fisherman.

Soon after the LTV landed on the bottom, lobsters smelled the bait and quickly found their way to the trap. If the trap's kitchen was unoccupied, more than half of those that approached entered and nibbled at the bait for about ten minutes. An astounding 94 percent of them walked right back out. Furthermore, while one lobster was eating, other lobsters were battling among themselves to be the next to enter, reducing the potential catch drastically—especially since the one eating also did his utmost to fight off the intruders between bites. In one twelve-hour period recorded on video, lobsters in the vicinity made 3,058 approaches to the LTV. But most of the approaches were repelled because of aggressive interactions with other lobsters.

Only forty-five lobsters succeeded in entering the trap, and of those, twenty-three ambled out one of the kitchen entrances after eating. Twenty prolonged their stay by entering the parlor, but seventeen of those eventually escaped, leaving just five in the trap. Of those five, three were under the legal size. When Win and his students hauled the trap up, they had caught a grand total of two salable lobsters.

What the LTV revealed about the inefficiency of the trap as a tool for catching lobsters spoke volumes to both Bob Steneck and Jack Merrill about why the lobster population might have endured despite all predictions to the contrary. But what really captured Bob's imagination was the constant disputes that were a feature of life in traps. Like a dysfunctional family, lobsters pushed each other around the kitchen and kicked each other out of the house. Other researchers had tried stocking a trap with several large lobsters and found that the presence of lobsters in the parlor alone discouraged new lobsters from entering, even without a fight. It seemed to Bob that a female trying to protect her eggs would have every reason to stay away, thus making her all but invisible to lobstermen—and so Bob's analogy to the *Star Trek* cloaking device.

A human parent might point out that protective behavior begins not with a baby's birth but with the onset of pregnancy. If female lobsters began avoiding traps before they even extruded their eggs, that might help solve the mystery of why V-notching worked as well as lobstermen claimed it did. Jack Merrill had never been able to explain how new eggers avoided traps long enough to mature, mate, and extrude eggs onto their tails so they could earn a V-notch. Indeed, according to the calculations of the scientists charged with managing the lobster population, a female lobster surrounded by traps had almost no chance of attaining motherhood. But as their ovaries matured and began to fill with eggs, those girls might steer clear of the packs of lobster toughs that hung around traps beating each other up. And Jack was right. For a brief period every fall, lobstermen caught eggers in droves. To Bob this didn't seem quite as magical as it did to Jack. Bob guessed the big females were simply fattening themselves up before winter. For a brief few weeks, the need to acquire nourishment might outweigh the risks.

Bob and Carl adjusted their sea-sampling schedule to include the autumn run of eggers. But the possibility that large numbers of reproductive females might be avoiding traps for much of the year pointed to another problem. Sea sampling generated useful new data. But even viewing the world through the eyes of lobstermen wasn't sufficient to show Bob what, as an ecologist, he most wanted to observe—the distribution and abundance of an organism in its habitat. There had to be large lobsters on the bottom that fishermen never saw. Bob needed to get down there and find them.

16

Victory Dance

The ship lay on the horizon ten miles southwest of Little Cranberry like a man-made island of white steel. The lobster boat, tiny by comparison, slid alongside the hulking craft. Bruce Fernald, Jack Merrill, and several local lobstermen leaped up across a gap of sea and were pulled aboard by members of the ship's crew. The fishermen had boarded the 170-foot R/V *Edwin Link,* a state-of-the-art research vessel that had come to Maine from the Harbor Branch Oceanographic Institution in Fort Pierce, Florida. It was Jack's second day aboard the *Edwin Link*, but Bruce's first.

Bruce gawked as he and Jack were led past a compressor room that housed gas-transfer pumps and banks of filtered scuba air, a maintenance lab where bins and drawers contained duplicate electronics and spare parts, and a machine shop where technicians could fabricate any metal device they needed. They walked past a wet lab with saltwater tanks and entered a three-hundred-square-foot dry lab, where Bob Steneck was hunched over a chart. He looked up when Bruce and Jack entered.

"Hey, guys!" Bob said, a grin bursting from under his orange beard.

"Quite a craft you've got here," Bruce said.

"Isn't it?" Bob responded. "Make yourself at home. We'll be ready shortly."

Bruce wandered past a comfortable lounge with couches, magazines, and satellite television. He found himself in a galley that could seat eighteen.

In tanks under Bruce's feet sloshed sixty-two thousand gallons of diesel fuel and forty thousand gallons of drinking water. The deck vibrated with a distant hum from electrical generators and a pair of 1,000-horsepower engines.

Bruce climbed a stairway to the bridge. The room offered a commanding view of the sea, but Bruce was more impressed with the view inside. The bridge was outfitted with two radar units, a magnetic compass, a gyrocompass, an automatic pilot system, a digital gyro repeater, satellite e-mail and fax, multiple satellite phones, video echo sounders, sonar, and a twin-scope scanning depth recorder—but no steering wheel. Bruce had been driving boats for more than twenty years and he was astonished. All 288 tons of the *Edwin Link* was steered with a tiny lever four inches long.

"If Bob is looking for lobsters," Bruce remarked, "he sure as hell got himself the ultimate lobster boat."

Bruce peered aft through the windows of the bridge, where an A-frame crane towered forty feet over the ship's stern. He stepped outside and descended two flights of steel stairs. There, poised under the crane and surrounded by a jungle of equipment, was the method of Bob's madness—the *Johnson Sea-Link*, a four-man submersible.

Bob had chartered subs before, but not for the task at hand today. Bob would be looking for brood-stock lobsters along the coast, and the *Johnson Sea-Link* was well suited for a visual search. The sub's forward command chamber was a transparent sphere of clear acrylic, with walls five inches thick, that could accommodate the sub's pilot and one scientist. Behind this orb, the body of the sub extended for twenty feet—a framework of metal tanks, ballast compartments, electric motors, and a second passenger chamber that was tubular and constructed of aluminum. The aft chamber could accommodate one technician and an additional researcher.

On the front of the sub, surrounding the clear acrylic orb, were directional thrusters, and underneath was a platform fitted with mesh buckets. Suction hoses fed an array of plastic containers for capturing sea creatures. From the bottom of the frame jutted a pair of tubular fangs. Peering at Bruce was a sort of giant insect eye with four little lenses surrounding a probing black iris—the sub's video camera. Reaching toward his head was a metallic claw that could have crushed his skull. Standing at the business end of the *Johnson Sea-Link*, Bruce mused, brought to mind a confrontation with a monstrous robotic lobster. He shuddered.

From the ship's superstructure Bob strode out on deck and was joined by the sub's pilot. Jack was close behind, along with several other fishermen. Bob scurried up a yellow ladder leaning against the sub and dropped into the clear sphere through a hatch on top. Relishing his position of command, Bob had decreed that one lucky lobsterman would be allowed to dive with him. Bob and the sub's pilot would sit in the forward command sphere, while a technician and the lobsterman would ride in the enclosed chamber aft.

The *Johnson Sea-Link* was capable of diving to three thousand feet. On this trip it would descend only a few hundred, but traveling even to those depths wasn't to be taken lightly. Each of the dives would last only a couple of hours, but the sub carried enough air for the occupants to stay alive for five days. The sub had a duplicate set of batteries, but in the event of a complete loss of power, manual carbon-dioxide scrubbers could be used to detoxify the air. When those lost efficacy, emergency breathing equipment would buy the occupants a little more time.

By now Jack was familiar with the drill. The previous day he had watched a lobsterman volunteer to join Bob on a dive. The fisherman was ushered to the sub's aft chamber, accessed through a hatch on the sub's underside. It was no transparent orb like the forward sphere, but a dark compartment of three-inch-thick metal walls with only a tiny porthole on each side. The man had glanced back at the surface of the sea, sparkling in the sunshine, then lifted himself into the chamber.

While Jack stood by, preparations for the dive had proceeded for ten minutes and then stalled. The lobsterman emerged from the sub. He was sweating.

"Sorry."

"Anybody else?" a technician shouted.

Jack stepped forward.

"I'll go," he said.

Jack hoisted himself inside. The tubular chamber was eight feet long and less than four feet across. He lay down on padding next to the submersible's dive technician, who explained a variety of procedures Jack must follow in case of emergency.

"If we have fire or smoke in the aft chamber," the technician said finally, "the first thing we're going to do is inform the pilot up front by yelling, 'Fire! Fire! Fire!' really loud so he can hear us. Then immediately we'll turn off our oxygen supply."

With that, the hatch beneath them was pushed shut and dogged tight.

The *Edwin Link* shuddered as the hydraulic pistons on either side of the A-frame groaned. The towering crane, holding fifteen tons of submersible plus another few tons of launch machinery, heaved off the deck, then pivoted out over the stern until the sub was dangling over the wash from the ship's propellers. From the center of the A-frame hung the launch mechanism. At its tip was a cone-shaped device four and a half inches in diameter, made of wedges of pure titanium. It was called the drop-lock, and it held the sub's entire weight. Gravity pulled the titanium wedges of the cone open like the petals of a flower, enlarging the diameter of the drop-lock so that it remained fixed inside a socket atop the sub.

A fist-thick cable played out, and the sub settled into the water in an eruption of foam. When the cable went slack a burst of compressed air forced the titanium wedges of the drop-lock together, and the cone came loose. The *Sea-Link* was free.

Jack could see nothing out the portholes except green light and bubbles, but he could feel the sub tossing on the waves. As the sub sank below the surface all movement subsided. The green outside darkened to blue and then black. An eerie silence filled the chamber, interrupted every few minutes as the pilot read the sub's depth into a voice transmitter. His words were broadcast back to the ship as though he were shouting through an underwater megaphone.

Xenon-arc floodlights clicked on outside. Jack peered at the forward video monitor but all he could make out was a rain of plankton. Then he saw it—the ocean floor, a pasty sediment spotted with boulders. Propellers whirred, the sub leveled, and from the pilot sphere Bob's voice indicated that he was commencing a transect. As the sub embarked on its journey across the bottom to hunt for lobsters, Jack glimpsed the tail of a fish, fleeing in a puff of silt. He spotted several sea urchins and caught sight of a crab, running sideways on tiptoe.

Jack smiled. It was a landscape he had never seen before, but after twenty-five years of hauling up bits and pieces of it, it looked awfully familiar.

~

A month later a group of lobstermen gathered with their families in the Grange hall on Little Cranberry Island, pulled down the window shades, and popped a videotape into the VCR. The children scooted into a clump in front of the screen.

The tape was from Bob Steneck and it showed what the *Johnson Sea-Link*'s video camera had recorded during the dive. One of the lobstermen explained that Dr. Steneck was a scientist, and that he was using a submarine to go down to the bottom of the ocean to count lobsters. By using science to count lobsters, Dr. Steneck was helping lobstermen learn if the ocean had enough mother and father lobsters. It was like—well, if your parents didn't live on Little Cranberry Island and have you as their children, then the island wouldn't have anybody living on it in the future. Of course, lobster mothers and fathers were different from people mothers and fathers. Lobster mothers had to work extra hard because they made so many babies—thousands of babies. How would you like to have five thousand brothers and sisters?

The video started and the ocean floor passed across the

screen. Then there were rocks, starfish, and seaweed. The audience was viewing the bottom of the sea, filmed a few miles from where they were sitting. From the underwater gloom, a huge lobster lumbered into view. The children squealed. The sub slowed and aimed measurement lasers at the approaching monster. Undeterred, the lobster strode toward the sub and raised its claws. The laser dots, about four inches apart, encompassed only three-quarters of one of the claws. The lobster walked directly toward the sub—its antennae lashing and its pincers wide. A claw struck out toward the camera, scattering the children in the front row. The lobster backed up a few paces, raised both claws over its head, and spun in a circle. One of the boys whooped.

"It's doing a victory dance!"

~

Whenever Bruce Fernald loaded traps aboard the *Double Trouble,* he tried to time the job to coincide with the high tide so he could back his truck down the wharf and heft the traps straight from the bed of the pickup onto the deck of the boat. He'd never thought he'd be doing the same thing with his personal belongings.

"Hang on, tide," he pleaded. The tide had already begun to ebb, and the *Double Trouble* had dropped a foot in the time he'd been back to the house for, what, the fifth load in the past three days? The dock was only half again as wide as his truck, so Bruce couldn't afford to rush. He'd driven backward down this wharf a thousand times, his view obstructed by six-foot piles of traps in the back, but he was still capable of running off the edge and dumping the twins' computer or Barb's packed-up jewelry studio into the drink.

"Really," Bruce told the ocean, "there's no need of this unnecessary bullshit."

Except for his years in boarding school and the navy, Bruce had lived on Little Cranberry Island his entire life. There had been many reasons to stay, but now he added another. He hated moving.

With the load stowed in the stern, Bruce, Barb, and the boys boarded the *Double Trouble* for the final run over to Mount Desert, where they unloaded the boat and jammed their belongings into the back of their station wagon. It was a three-minute drive up the hill to their winter rental. They parked and hauled everything inside. Bruce and Barb stood among the boxes and looked at each other. After eighteen years in the snug white-and-blue house they'd built together and called home, they were no longer living on the island.

Back in June, Warren and Ann had joined Bruce and Barb and the rest of the Little Cranberry community in the Grange hall for graduation. The island school didn't have eighth-graders every year, so it was a special moment when three boys, two of them Bruce and Barb's twins, stood on stage and received the accolades of their teacher, their families, and the seventy or eighty other villagers who'd helped raise them. Warren had stood on that same stage when he'd graduated from the island school. It wasn't hard for Warren and Ann to conjure up the day that Bruce had received his eighth-grade diploma in that same hall. A few months later, Bruce would become their first child to leave home.

But Bruce, Barb, and the twins had ruled out boarding school. They wanted to stay together. Yet the options were limited for island children after eighth grade. The twins could either commute to the high school on Mount Desert Island aboard the ferry, forfeiting the chance to participate in extracurricular activities in the afternoons and evenings, or the whole family could move off Little Cranberry from September through May each year until the twins left for college. Bruce had calculated that he could afford nine months of rent a year if the lobster catch remained as lucrative as it had recently become.

It had seemed like a clever plan, until now.

"Damn!" Bruce cussed. He shook his head, then laughed at his lack of foresight. Carrying another box up the stairs, he had just realized he'd be moving seven more times in the next four years.

The summer sailboats had been hauled onto land, and the pleasure yachts had left for warmer climes. Bruce secured a vacated harbor slip for the *Double Trouble* and began leaving at dawn every day from Mount Desert. He swung by Little Cranberry in the morning to pick up his sternman, and again in the afternoon to sell his catch and fill his bait bin. Not living on the island felt strange.

Once he was hauling his first trap of the day, though, Bruce no longer had time to think about it. That autumn, the wave of high catches that had begun mysteriously in the western half of the state rolled into the area around Mount Desert Island with full force.

Lobster mania hit Maine like a gold rush. Distant relatives of lobstermen suddenly wanted to buy boats. Groundfish draggers who'd given up on cod, haddock, and hake traded in their rusting trawlers for shiny lobster boats and muscled their way into coastal territory. The number of traps in Maine waters hit 2.8 million, and in western Maine you could practically walk across the buoys, they were so thick. Fights broke out between men who'd tangled each other's gear into knots. The job of sternman became lucrative enough that there were waiting lists.

On Little Cranberry Island, Jack Merrill's and Dan Fernald's houses were next door to each other, and Jack's twelve-year-old son, in too much of a rush to wait for his dad's help, dragged a couple of Dan's traps down to the dock by mistake and set them in the harbor. Aboard the *Bottom Dollar,* Jack was in too much of a rush to gaff his next buoy one day, and punched the throttle at the same moment he noticed a loop of rope wrapped around his ankle. As the boat lurched forward Jack was yanked aft like a marionette. Shouting in pain, Jack braced himself against the transom while his sternman rushed to the controls to stop the boat. Fortunately, the rope pulled Jack's boot off instead of dragging him underwater. He was left with a badly torn knee.

Closer to land, residents of Little Cranberry who'd never lobstered before started setting traps. After a day of sorting mail, Joy, the postmistress, whose father had been killed lobstering in a storm when she was a girl, swapped her miniskirts for jeans and a pair of rubber boots and set out to haul a few traps from a skiff. Soos, the lady who owned the island's general store, took time off from sewing teddy bears and taking inventory in order to set traps from a fiberglass dinghy.

Stefanie, who had worked as a sternman for her husband for years, bought herself a lobster boat, named it after her daughter and mother, and became the first female fishing captain in the history of the island. She hired a college girl to be her sternman for the summer.

By the year 2000, following half a century of catches that had hardly wavered, Maine's haul of lobsters had tripled in just over a decade. Still, no one knew exactly why. The American lobster was now the most valuable marine species in the northeastern United States. Two-thirds of all lobsters caught in New England were coming from Maine.

That July, just as the Little Cranberry lobstermen began hauling in the earliest and biggest run of shedders they'd ever seen, another committee of government scientists issued a new assessment of the lobster stock. The document was five hundred pages long. This time it included a minority report written by Bob Steneck, but again the assessment's official conclusion was that the lobster population was being overfished. Not enough lobsters were getting the chance to mate and produce eggs before being caught. The scientists advanced a plan to rebuild egg production. To achieve this goal, they once again advocated raising the minimum legal size for lobster in the state of Maine.

Even Bob Steneck, in his letter to the editor, had once written that "increasing the minimum size might be a prudent thing to do." Perhaps that was still true, despite all the eggers and notchers that lobstermen had protected. Waiting for lobsters to grow bigger before hauling them up for sale might even be lucrative. The problem with the idea was the same as it had always been. Lobstermen didn't know whether consumers were flush enough to shell out the extra money for bigger lobsters.

Yet changes in consumer preferences since the 1980s had presented the lobster industry with a new marketing opportunity. Thanks to conservation by lobstermen, their fishery had produced sustainable harvests for half a century while other fisheries had stripped the seas. American consumers had learned to value—and pay extra for—such products as organically grown vegetables, free-range chicken, and wild salmon caught with hook and line. Surely, those same consumers would purchase a slightly larger lobster that was promoted as an additional contribution to the sustainability of harvests. In a sense, Maine lobstermen had become environmentalists well before such things were fashionable, and now that legacy might earn them dividends in ways old-timers like Warren Fernald had never imagined.

Still hauling traps aboard the *Mother Ann* at age seventy-four, Warren Fernald took the government's latest accusation of overfishing as one more insult. But at the same time, Warren felt his sons and their friends could afford to sacrifice a little for the lobster population. Warren had never fished more than four hundred traps at a time, and he'd been pleasantly surprised at how well the first increase in the minimum size had turned out. After fifty years of hauling, what he saw now was a fishery getting out of hand. There was a difference between chasing a gold rush and living the good life. Watching his sons work themselves to the bone, Warren worried that the generation he'd nurtured might have forgotten what it was they were fishing for.

"The way I fished, I always had fun," Warren would say after a short day on the water, easing into the armchair he'd worn bare after thirty years. "And I never lost an ounce off my belly from starvation."

Bob Steneck retrieved a manila envelope from his mailbox and frowned. There was no return address. He ripped it open and found an unmarked Zip disk inside. Popping the disk into his computer, Bob clicked his mouse on the icon and opened it.

"What the—?"

Someone at the National Marine Fisheries Service had leaked Bob the *Albatross* data. The keyboard clicked rapidly as Bob typed in a string of instructions to query the database. The results popped up, and he leaned back in his chair and let out a whistle.

"You have *got* to be kidding me."

For the past decade, the number of large, sexually mature female lobsters that had been caught in the net of the R/V *Albatross* had gone not down but steadily up, and this in spite of the limitations of trawling. The number of large males had also risen.

"No wonder they didn't want to give it to me," Bob muttered. "It says the same thing I've been saying." He printed out a graph of the *Albatross* catch and tacked it like a trophy to the door of his lab.

So far, the data from sea sampling and submersible dives—and the government's own trawl survey—indicated that more egg-producing lobsters roamed the bottom than government scientists had been willing to admit. The data also indicated that the practice of V-notching was more effective than the government had supposed—75 percent of all egg-bearing lobsters observed on sea-sampling trips already carried the notch. Surely, without enough eggs, the lobster catch could hardly have tripled in size.

But if there were enough eggs, what had caused the numbers of superlobsters and babies to plummet, even as the catch rose? The ecologists returned to their original question.

Rick Wahle had come back to Maine in 1995 and, like Lew Incze, was now a research scientist at the Bigelow Laboratory for Ocean Sciences. From the window of Rick's new office in Boothbay Harbor he could look across the water at Damariscove Island on the horizon. The view dogged him because the low settlement he'd seen at the nursery off Damariscove's western shore hadn't improved.

By the end of the 1990s, Rick and Lew had witnessed half a decade of declines in the number of superlobsters and baby lobsters, and that seemed sure to slash the future population of adults. Rick and Lew had mentioned the decline at a forum for fishermen, but they had not been able to say when, where, or how catches might be affected, and with catches at an all-time high, few lobstermen had cared.

Making an accurate projection would require the use of sophisticated mathematical forecasting techniques. The calculations would be complex because not all lobsters grow at the same speed. Dominant lobsters acquire more food and molt more frequently than subordinates, and when lobsters leave the nurseries, some of them encounter warm water and grow quickly while others end up in cold water and grow more slowly.

For assistance with these calculations, Rick and Lew turned to a biologist named Michael Fogarty. A former student of Stan Cobb's at the University of Rhode Island, Mike was an expert on lobster-population dynamics. For years, Mike had been designing mathematical models to simulate the growth and development of lobsters. Much of his work had demonstrated the benefits of raising the minimum legal size.

Now Mike agreed to develop a different model, one that would use Rick and Lew's data to predict future lobster harvests. Rick, Lew, and Mike presented the initial results of their collaboration at an international conference on lobster science in Key West, Florida, in the fall of 2000, just weeks after the government's latest stock assessment. Like the government's report, the paper presented by the three scientists was pessimistic. The announcement created a stir in scientific circles.

Back at the University of Maine, Bob Steneck combed through the data that he and Carl Wilson had been collecting during their scuba dives along the coast. Bob detected a corresponding drop in juvenile lobsters. Though catches were still strong, all other signs were ominous.

Most lobstermen didn't monitor arcane conference proceedings in Key West. That winter Rick, Lew, and Bob debated whether to inform the public in Maine. Their system had yet to be proven predictive, but in January of 2001 they issued a joint press release warning of a possible decline in lobster abundance.

The ecologists explained that the implications for lobstermen were still unclear, but that three independent sets of data—superlobsters, babies, and juveniles—suggested that catches could fall in western and midcoast Maine sometime in the next few years. Bob was particularly worried because the decline in juveniles appeared to encompass what had become the state's richest lobstering grounds—the area from Pemaquid in the west to Mount Desert in the east.

Within days the scientists' announcement was being reported in local newspapers. With catches higher than ever, the reaction of lobstermen was mixed—about one part acceptance, one part respect, and ninety-eight parts disbelief. Lobstermen were surprised that a statement from their friend Bob Steneck sounded similar to what government scientists had been saying for decades.

But the viewpoint of Bob and his colleagues was actually quite different. Government scientists had been arguing that low egg production could lead to a collapse of the lobster population. Rick, Lew, and Bob were suggesting that the decline had nothing to do with egg production. Sea sampling, submersible dives, and even the federal trawl survey all indicated that the population of large egg-producing lobsters was sufficient—perhaps even at a surplus.

The problem, the ecologists thought, was that not enough of those eggs were becoming lobsters. Lew's data showed that fewer larvae than before were arriving at the nursery grounds. The question was why.

17

Fickle Seas

Ten days after Rick Wahle, Lew Incze, and Mike Fogarty issued their prediction of a lobster decline at a scientific conference in Key West, a Titan II intercontinental ballistic missile blasted off from Vandenberg Air Force Base in California. Traveling at seven thousand feet per second, it was thirty-eight miles up when the rocket's first stage fell away in a burst of flame. The second stage ignited, blasting the payload even higher, and then seven minutes into the flight it detached. For decades the missile had been tipped with a 9.6-megaton nuclear warhead aimed at the Soviet Union. Now it carried a different payload.

The Titan II had just lofted into space a satellite called NOAA-16. The satellite fired its own rocket and was soon orbiting more than five hundred miles above the earth. Its orbit was polar, which meant that NOAA-16 passed almost directly over the North and South Poles on a fixed trajectory while the earth rotated beneath it. NASA's Goddard Space Flight Center in Greenbelt, Maryland, took command of the satellite.

On board, an elliptical mirror the size of a dinner plate starting spinning, looking at a swath of the earth fifteen hundred miles wide and reflecting thermal infrared radiation into a collimating telescope called the AVHRR—Advanced Very High Resolution Radiometer. Once the satellite had completed initial testing, its command was transferred to the owner, the U.S. National Oceanic and Atmospheric Administration. Circling the earth fourteen times a day, NOAA-16 began collecting massive quantities of data, including sea-surface temperatures for every square mile of the Gulf of Maine.

Also circling in the earth's polar orbit was QuikSCAT, a NASA satellite launched on a Titan II from Vandenberg the previous year. QuikSCAT carried a rotating parabolic dish that wiped twin-beam microwave pulses in a thousand-mile circle across the sea every three seconds. The dish, part of a device called a scatterometer, detected echo and backscatter in the return signal and routed this radio-frequency energy through rectangular pipes to an onboard computer called the SeaWinds Electronics Subsystem. From the scatter signal the computer reconstructed the size, shape, and orientation of ocean waves and converted that information into the velocity and direction of sea-surface winds across 90 percent of the globe.

Two months before NOAA-16 hit the skies, a robot floating off the Maine coast had automatically dialed its new cell phone for the first time and begun transmitting data. Painted yellow and sporting four solar panels and an antenna, the tethered robot recorded hourly data on wind speed and direction, wave action, visibility, and irradiance. Underwater it trailed a feeler 160 feet deep that measured the velocity and direction of currents at several depths while also tracking temperature, salinity, turbidity, oxygen content, and chlorophyll concentration.

With funding from the Office of Naval Research, six more robots were deployed in rapid succession off the Maine coast the following summer, including one stationed twenty miles southeast of Little Cranberry Island, near Mount Desert Rock. These were soon followed by three more, all them phoning in regularly to report their data. Still in the planning stage were four pairs of antenna arrays along the edge of the Gulf of Maine called CODAR—Coastal Ocean Dynamics Application Radar—to map surface currents twenty-four hours a day from one end of the gulf to the other.

The NOAA-16 AVHRR temperature grids and data from similar satellites, the QuikSCAT sea-wind plots, and the tethered robot readings—along with the planned addition of CODAR maps—together form the most ambitious oceanographic data-collection effort ever attempted. In combination they are called GoMOOS, the Gulf of Maine Ocean Observing System. The GoMOOS data converge at a satellite receiving station at the University of Maine and are streamed into a circulation-modeling computer.

Every morning at two o'clock the computer queries the National Oceanic and Atmospheric Administration for updated information on weather conditions. It downloads wind- and air-pressure data from the National Centers for Environmental Prediction, and talks to the yellow robots on their cell phones from their lonely posts off the coast. The computer uses this information to build a model of temperature, salinity, and current velocity throughout the Gulf of Maine. The output, nearly in real time, is three-dimensional color-coded images of the gulf, with the direction and speed of ocean currents indicated by a layered grid of curving arrows on three planes. Overlays of temperature and surface-wind satellite imagery refine the picture further.

GoMOOS was conceived by a small group of oceanographers as a joint project of the University of Maine and the Bigelow Laboratory for Ocean Sciences. One of the oceanographers was Lew Incze. When GoMOOS came online, it became one of Lew's most important tools in the quest to figure out where lobster larvae, creatures the size of ants, were being carried across the thirty-six thousand square miles of the Gulf of Maine. There was still no substitute for observing actual animals in the ocean, but Lew had come a long way from towing a two-foot-wide net behind a boat.

Working with GoMOOS was a little like playing Poseidon. Lew could peer down from heaven, reach out a hand, and peel away a layer or two of ocean. What he saw didn't so much bestow omniscience as humble him. The sea's movements were staggeringly complex.

~

The Gulf of Maine is essentially a large bowl. On the bottom are three separate basins, two of them nearly seven hundred feet deep. The third—Georges Basin, on the gulf's seaward side—is more than a thousand feet deep. The gulf's western rim is formed by land, from Cape Cod up to Maine and over to Nova Scotia. But in the east, the gulf is rimmed by underwater banks that rise nearly to the surface, including Nantucket Shoals, Georges Bank, Browns Bank, and the Scotian Shelf. This part of the gulf's rim has an indentation that allows deep water to flow in from the Atlantic Ocean and back out. The indentation, called the Northeast Channel, is only twenty miles wide, but it goes down more than seven hundred feet. A much shallower spout, called Great South Channel, allows limited flow in and out of the gulf between Georges Bank and Nantucket.

Water flow inside the bowl is dominated by a huge, gulfwide gyre that constantly circles counterclockwise along the rim. Water from the North Atlantic enters the gulf across the Scotian Shelf and along the northern wall of the Northeast Channel and flows up the coast of Nova Scotia, past the mouth of the Bay of Fundy, and southwest along the Maine coast. The northwest leg of this gyre is a rapidly moving plume of cold water that oceanographers call the Eastern Maine Coastal Current.

About two-thirds of the way down the Maine coast, much of the water in the Eastern Maine Coastal Current is deflected away from shore into the interior of the gulf. The rest continues down the coast in a warmer, slower plume called the Western Maine Coastal Current. Off Cape Cod, a small amount of this water exits the bowl through the Great South Channel, but most of it turns east and then north, flowing back up the inside edge of Georges Bank. Some of the water that was deflected from the Eastern Maine Coastal Current rejoins the gyre here.

The gyre flows northward and returns to the Northeast Channel, where much of the water exits the gulf and flows back into the North Atlantic. Some of it remains inside the gulf to cycle through the gyre again. The currents inside the gulf are chaotic, and a myriad of eddies and vortices complicate their movements. Oceanographers have calculated that an average parcel of water spends about one year traveling inside the gulf before it leaves.

When the lobster hatching season begins, usually between mid-June and early July, large numbers of females carrying fully developed eggs undergo abrupt contractions of their tail muscles during the night. Over the course of a week or so, these nocturnal contractions shake each lobster's thousands of embryos free. The embryos are soft and round when they break through the outer seal of the egg, but within a few minutes they assume the shape of larvae, pointy-tailed and shrimplike.

A first-stage lobster larva can detect the gravitational pull of the earth and swims upward and away from the bottom, beating its paddlelike appendages furiously. It can also detect light, and during the day it swims toward the sun. These instinctive behaviors deliver the larva to the surface, where it comes within reach of the wind. The larva slides along with the topmost layer of water, skimming the sea with the breeze.

But the larva's foray at the surface is short-lived. After several days the first-stage larva sheds its shell and enters a second stage, which is more sophisticated. It has grown only a millimeter but it has gained new appendages and muscles. It floats below the surface, perhaps ten feet under or even deeper, putting it at the mercy of the gulf's complex currents. Soon it molts again and becomes a third-stage larva. It remains low in the water, continuing to travel with the currents. By now it has developed many of the characteristics of a lobster, including tiny claws, swimmerets, and tail flippers. Finally it becomes a postlarva, or superlobster, and returns to the surface to swim and sail on the wind. After a few days the superlobster begins to dive, searching for cobblestones in shallow water.

Lew Incze's specialty was larval ecology, and he knew that it could take a lobster larva anywhere from twenty to forty days, sometimes even longer, to develop from a hatchling on its mother's tail to a baby bedded down in a nursery. Given the way wind and water moved in the gulf, that was enough time for a larva to travel quite a distance. Lew teamed up with a physical oceanographer who combined AVHRR satellite readings and other data with a state-of-the-art circulation model running on a computer. The model would divide the gulf into thousands of three-dimensional triangles and compute the effects of tidal transport, temperature, salinity, and turbulence, along with the gulf's dominating counterclockwise gyre. Lew could then track individual particles moving inside the flow field.

Lew integrated a biological model of the lobster's larval life cycle into his colleague's physical model of the ocean. Coupling biology to physics was a delicate art. The biological model had to simulate an individual larva hatching and then developing through all three larval stages prior to the superlobster stage. Lew could then assign a hatching location to a hypothetical larva and run the biological model inside the physical model to see where the larva ended up in the ocean when it was ready to become a superlobster. Alternatively, Lew could run the model backward. He could assign an arrival location for the larva and, as if he were pressing Rewind on a video, tell the computer to run the currents in reverse to calculate where the larva might have hatched.

Using this reverse method, Lew chose a series of end points around the Pemaquid region of Maine's western coast, indicating where the hypothetical larvae had arrived. By running the model backward, he might get an idea of where the superlobsters that he and Rick had been counting for the past decade had been coming from.

For some of his arrival locations, Lew selected the coastal nurseries. But it was possible that superlobsters also arrived from farther offshore. The physical model indicated that even a weak onshore breeze could propel a superlobster into the nurseries from fifteen or twenty miles out. So in addition to the arrival locations near shore, Lew selected a string of locations that were between fifteen and twenty miles out to sea.

When Lew ran the model backward, the computer drew a series of white lines onto a map of the Maine coast, leading back in the direction from whence the currents would have come. Each line covered a series of upstream locations where the virtual larva could originally have hatched from under its mother's tail.

The results fell into two rough categories—larvae that traveled short distances and larvae that came from farther away. The computer calculated that some of the larvae in the nurseries had probably hatched nearby, the white lines forming short squiggles that corresponded to local currents. That made sense. Female lobsters that stayed in shallow water might not be exposed to the larger currents in the gulf, and their offspring wouldn't have traveled far. Lew supposed that local larvae, traveling short distances close to shore, might account for much of the settlement in the nurseries in the central and western half of the Maine coast.

But for the larvae that Lew had programmed to sail into the nurseries from offshore, the white lines that the computer drew looked quite different. They weren't short squiggles. They were long-distance highways. They stretched up the coast for a hundred miles or more, the hatching locations well into eastern Maine and beyond. These virtual larvae were riding the Gulf of Maine's counterclockwise gyre down the coast, rafting the powerful rapids of the Eastern Maine Coastal Current.

Lew suspected that each of these highways had on-ramps along its entire length, with larvae joining the traffic from any number of locations off Nova Scotia, the Bay of Fundy, and Down East Maine. It was possible that the delivery of larvae over long distances had helped fuel the increase in catches along the western half of the Maine coast during the 1990s.

To the extent that long-distance larvae might have supplemented local larvae in western Maine, Lew thought it likely that they had come from a variety of locations throughout the northwest gulf. All the same, during the computer's rudimentary simulation it was remarkable how many of the white lines converged in the vicinity of a single large island just over the Canadian border called Grand Manan.

Dan and Katy Fernald's daughter Erin loved Wellesley College, but she also loved Little Cranberry Island. Dan and Katy had faced some opprobrium for pushing Erin to seek a highbrow education, but now Erin enjoyed the best of both worlds. As her sophomore year at Wellesley drew to a close in the spring of 2001, she knew she would miss the stimulation of attending classes, but she was looking forward to returning to the island to work in her parents' art gallery for the summer.

But she wasn't going anywhere until she wrote her term paper for Econ. 228—Environmental Economics. Erin had grown up listening to her dad, her uncles, and Jack complain about the government's approach to managing the lobster fishery. She'd also heard stories about her mother's college thesis on the economics of lobstering. Now that Erin had the resources of a world-class educational institution at her disposal, she thought it might be time to form her own opinion.

The term paper had been assigned as a team project. Erin broached the idea of writing about lobsters to her team partner, a student from California, and was surprised when she readily agreed. A string of long nights in the library resulted in a twenty-page monograph titled "Resource Allocation and Regulation of a Common Pool Resource: The Lobster Industry of Maine."

On the last page of the paper, Erin and her classmate wrote that "managing a fishery is a complex process, involving economics, biology, and the study of the social climate of the fishermen themselves." Having grown up on Little Cranberry Island, Erin thought she had a handle on the social climate part. Economics too made more sense after taking the class and researching the paper. But she was still curious about biology. Before leaving for summer vacation, Erin decided to sign up for a biology class when she returned to campus in the fall. She had been thinking of declaring a major in one of the sciences.

Back on the island, Jack Merrill got wind of Erin's new interest in lobster biology and mentioned that Bob Steneck was planning another research cruise to study lobsters. He would be passing right by Little Cranberry.

A week later Erin was standing on the deck of the R/V *Connecticut,* where she was introduced to Bob and his crew. She also made the acquaintance of the *Phantom,* the underwater robot that was Bob's latest tool in his quest for large lobsters. Now that he had scoped out the seafloor using the *Johnson Sea-Link* manned submersible, he could conduct quicker and cheaper follow-up surveys with remotely operated vehicles—ROVs—like the *Phantom*.

Erin's workday aboard the *Connecticut* began at 7:00 A.M. and continued at a frenetic pace until eight or nine each night. The *Phantom* completed five dives a day, generating an hour of videotape on each dive. During the dives, two research assistants sat in the darkened control room, glued to the *Phantom*'s live video feed from the ocean floor. One assistant typed commands into a computer that controlled the zoom angle on the *Phantom*'s camera while the other recorded depth, temperature, and time. After each dive a third assistant reviewed the videotape in the *Connecticut*'s lab. Each time a lobster appeared, the assistant hit Pause and gauged the lobster's size by the distance between the red dots projected by the *Phantom*'s lasers. An hour-long dive might uncover as many as forty lobsters.

Erin performed all of these tasks and learned to sleep next to the roar of the ship's 800-horsepower engine. The *Connecticut* operated twenty-four hours a day. Before Erin and the other dive assistants bedded down each night, their bunks were vacated by the *Connecticut*'s second shift. The night crew included a graduate student of Bob's who was working with Lew Incze on lobster larvae. From dusk till dawn he towed a fine mesh net behind the *Connecticut* at a variety of depths. In the morning Erin would open the refrigerator in the lab to find vials of somersaulting lobster larvae and superlobsters tapping at the walls of their containers.

Half the fun of the cruise for Erin was listening to Bob talk. In the darkened control room, with mud and rocks passing across the video screen as the researchers waited to find the next lobster, Bob would sip from his fifth or six cup of coffee and pontificate on evolutionary theory, phylogenetic classification, and the philosophy of science.

"I was thinking of maybe majoring in geology," Erin told Bob.

"Hey, that's what I majored in!" Bob exclaimed. "Actually, geology and biology—a double major. You should!"

As the *Phantom* traversed the Gulf of Maine, Erin saw all manner of sea life, including flounder, hake, cod, and ocean pout, along with baby octopuses, orange sea anemones, patchwork fields of black and blue sand dollars, and intricate coral formations. One day the *Phantom* even found a shipwreck. One night the *Connecticut* sailed through a migration of bats. Erin woke to find the winged mammals flying around inside the ship.

When the *Connecticut* passed into Canadian waters, Erin witnessed the one thing she most wished her father, her uncles, Jack, and every other lobsterman on Little Cranberry could behold—the seafloor off the island of Grand Manan. At the mouth of the Bay of Fundy, where the highest tides in the world rose and fell twice a day, the *Phantom* was lowered off the stern, and Bob gathered his crew around the video monitors in the command module.

"This," Bob said, "you have got to see."

The *Phantom* descended, and in a few minutes Erin could make out a murky plain of mud. The *Phantom* had gone only a short distance when a circular depression appeared, a dish five feet wide. Hunkered in the center was the biggest lobster Erin had ever seen. It was a female, at least two feet long, and from her tail hung perhaps a hundred thousand eggs. A few yards beyond was another pit with another mammoth mother, eggs bursting from under her tail. After that there was another, and after that, yet another. The bottom off Grand Manan was a vast expanse of egg-bearing lobster dens, one of the greatest aggregations of fecund females that had been found in the Gulf of Maine.

While Bob Steneck stalked females Down East with his underwater robot, Diane Cowan was clambering aboard lobster boats in the western half of the Maine coast with different technology. Her tools for stalking female lobsters were tubes of superglue and rolls of olive-drab duct tape. She also carried with her a container of yellow disks, each the size of a wristwatch, and a tray of solid white rods that looked like glue sticks.

Since founding the Lobster Conservancy, Diane had continued to count baby lobsters fourteen times a year at low tide with a zeal that verged on the religious. Local newspapers now referred to her as the Jane Goodall of lobsters, and the local lobstermen had come to expect her in the mud at water's edge at 5:00 A.M. or 6:00 P.M. or God knew what other hour. Soon the lobstermen were looking out for Diane like a friend, and they joked with her the way they joked with each other.

"How long you been turning over rocks looking for them baby lobsters, anyhow?" one fisherman asked.

Diane had to think for a second. "I guess it's been nine years."

"*Nine* years?" He laughed. "Jeez, you ain't as smart as I thought you were, are you?"

Diane had recently been hired as the state of Maine's chief lobster biologist, but astonishingly, even that job had presented the same problem as some of the teaching jobs she'd tried—too much time in an office, not enough time with the lobsters. So she'd quit.

Now she was living on an island with a winter population of three, in a small house overlooking an old lobster pound that had been donated to the conservancy. Backed by acres of woods, the house sat on a windswept ledge by the ocean. Her electricity came from solar panels and her heat from a wood stove. For half the year she hauled her water from an outdoor well. She'd relinquished day-to-day management of the conservancy to a staff on the mainland, who now oversaw eighty volunteers sampling baby lobsters at thirty sites around New England, including Little Cranberry Island.

Free again to focus on research, Diane hoped to use the cordoned-off cove that formed the old pound to rekindle her first love, the study of lobster sex. Her plan was to outfit the bottom of the cove with lobster homes spacious enough for two, rig up an underwater video surveillance system, and populate the cove with the largest male and female lobsters she could find.

In the meantime, Diane had devised a project to help answer the crucial question of where lobster larvae were hatching in the wild. While Bob Steneck and other scientists studied egg-producing lobsters in deep water using submersibles, ROVs, and other tools, Diane convinced fourteen of her fishermen friends to help her follow female lobsters in waters closer to shore.

On a typical autumn day Diane boarded a lobster boat at 5:00 A.M., her bag loaded with her collection of specialized equipment. When the fisherman hauled up an egg-bearing lobster, Diane noted the developmental state of the eggs and recorded the lobster's location using a GPS receiver. That was one way to determine where a female lobster might hatch her spawn. Another was to attach a homing beacon to the lobster's shell and track her.

Aboard the fishing boats, Diane deftly refashioned nearly two hundred new mothers. After drying an egger's back with a towel she wove a strip of duct tape—olive-drab to match the lobster's natural camouflage—through her fingers and squirted on three lines of superglue. Then she picked up one of her white rods, laid it on the center line of glue, wrapped the edges of the tape down around the rod to create a pair of sticky flaps, and attached the device to the lobster's back. The rod was a sonar transmitter that would emit a unique sequence of beeps for the next twelve months.

Next she tied a plastic bracelet around the narrow wrist behind the lobster's claw and attached one of her wristwatch-sized yellow disks. The disk was an automated thermometer that would record water temperature every hour for up to four years. Ready to go overboard, the lobster looked like a muscle-bound scuba aficionado just back from the dive shop—new tank on her back, snazzy yellow sports chronometer bulging on her bicep.

The fourteen lobstermen outfitted their fishing boats with hydrophones that could detect each female's sonar signal from half a mile away. While hauling traps, the fishermen, even if they never caught the lobsters again, could listen for them under the waves. Diane equipped her own open-decked skiff with hydrophones and motored into the bay at every opportunity to listen for her mothers through her headphones, each lobster clicking away on the bottom with one of thirty numeric codes on one of ten different frequencies.

Diane's task was complicated further when her Belgian sheepdog, Bear, insisted on going to sea with her. Though Diane was his fifth owner, he defended her from any threat as though she were his first. Threats, it turned out, were everywhere, especially aboard Diane's boat. When she spun the steering wheel a certain way, the growling outboard motor on the stern would turn to glare at Bear, and the dog would leap through the tangle of hydrophone wires to bite off another piece of the engine's rubber gasket.

Eighty percent of the sonar lobsters Diane released were successfully tracked or recaptured. Diane was startled to learn where they'd been. Many scientists have assumed that the first priority for an egg-bearing lobster is to find temperate water, where her eggs can develop quickly and hatch. But the movements of Diane's lobsters, and the logs of their temperature recorders, provided little evidence of such a search. The distances the lobsters traveled ranged from a few feet to a hundred miles. About a third of the females remained in the immediate neighborhood to hatch their eggs, about a third roved within a twenty-mile radius to find a hatching location, and about a third set off with their eggs over much longer distances.

If these movements were typical of egg-bearing lobsters everywhere, Diane guessed, she might have stumbled onto a reproductive strategy directly opposed to the well-known reproductive strategy of the salmon. Salmon return to their exact birthplace to hatch their eggs every generation, which leaves them vulnerable to natural or man-made shifts in the environment. By contrast, female lobsters as a group appeared to be hedging their bets by fanning out. To Diane, hatching eggs in as many places as possible seemed a wise strategy for a species that cast its young into the fickle currents of the ocean.

~

In another ocean-modeling session, Lew Incze ran the larval-transport simulation for the Gulf of Maine forward instead of backward. He gave the computer ten combinations of nearshore and offshore hatching locations around the gulf to mimic geographical variations in lobster migration, and he posited early, middle, and late hatching times for a hypothetical spawning season.

The computer's calculations showed that most of the larvae followed the gulf's counterclockwise gyre. As Lew had expected, many were delivered by the Eastern Maine Coastal Current from locations off Nova Scotia, the Bay of Fundy, and Down East Maine over long distances to the nurseries along the western half of the Maine coast. Not surprisingly, other larvae that had hatched near shore, especially in western Maine, were often not exposed to the gulf's large-scale currents and traveled only short distances, landing in local nurseries.

But what took Lew's breath away was the extent to which small changes in location, timing, and temperature at hatching could swing larval trajectories away from the nurseries, regardless of whether water temperature at the nurseries was hospitable. The computer simulations he was running were rudimentary, but the general implications were all too clear.

Under favorable conditions, any given nursery could receive vast numbers of larvae, many of them coming from a few miles away and others coming from a hundred miles up the coast. But under slightly different conditions, the larvae from distant hatching locations in the north, as well as the larvae that hatched near shore in the southwest, could be diverted by the gulf's powerful currents into deep and cold water offshore, leaving the coastal nurseries all but bare.

The American lobster is not unique in casting its young into the currents of the sea. Creatures as various as the codfish and the rock crab do the same thing. The difference is that when a female lobster tosses her hatchlings from her tail, she is, in a sense, aiming for a target. Codfish and rock crabs don't have nursery grounds, because their larvae aren't nearly as sophisticated as a superlobster. The larvae of the cod and crab are passive creatures that settle wherever the ocean puts them, which often means into the mouth of the nearest predator. By contrast, the superlobster's ability to seek out hiding places is the lobster's secret weapon. By exerting a degree of control over its fate, the superlobster vastly improves its chances of survival.

But this is also the lobster's greatest reproductive liability. A single cod or crab mother makes millions of eggs. For a mother lobster, the extra resources required to build her miniature superheroes means she can make only thousands. It is a risky strategy, because the delivery system that lobsters depend on—ocean currents—can fail to carry their limited numbers of offspring to their targets—the nurseries.

A variety of climatic and oceanographic phenomena can influence the currents from one year to the next, as Lew knew from monitoring the satellite, surface, and undersea data collected by GoMOOS. Shifts in prevailing winds, the orientation of the jet stream, cloud cover, and even the amount of ice melting in the Arctic could all affect how water moved around the Gulf of Maine. Any combination of effects was possible. A given nursery could experience both the retention of local larvae and the delivery of distant larvae. Or it could experience one without the other. Or, in the worst case, it could experience neither.

Because of climatic and oceanographic conditions inside the gulf, it was conceivable that an entire region of the coast could explode with baby lobsters or slump into vacancy, regardless of fluctuations in the number of female lobsters producing eggs. Lew suspected that a large portion of the larval supply to

Maine's coastal nurseries was locally hatched, but it seemed plausible that currents, temperatures, and other oceanographic factors along the coast had been preventing larvae—from both local and distant sources—from reaching the nurseries of western Maine during the second half of the 1990s.

Yet powers beyond the Gulf of Maine were at work too. By now Rick Wahle had more than a decade of data on baby lobsters, not just from western Maine but also from Rhode Island. Lew and Rick pored over the annual censuses from the lobster nurseries. The drop in baby lobster abundance from 1995 onward was similar in both locations. It was an astonishing fact, since oceanographic conditions in the Gulf of Maine and Rhode Island Sound are almost completely unrelated.

Lew concluded that some large-scale shift in a prevailing atmospheric system, in addition to currents specific to the Gulf of Maine, could be driving lobster abundance. There was an obvious candidate, though its influence remained unproven—the North Atlantic Oscillation. An eastern counterpart to El Niño, the oscillation is a titanic seesaw in pressure distributions over the North Atlantic that tips into a subtropical high or a polar low for years or even decades at a time.

The water that enters the Gulf of Maine tends to be dominated by either southern water from the mid-Atlantic latitudes, which is warmer, or northern water from the Labrador Sea, which is colder. The North Atlantic Oscillation can push the Gulf Stream away from the edge of the continental shelf, which appears to affect which type of water is dominant in the Gulf of Maine. Because the water inside the gulf has its own patterns of circulation, however, the connection is by no means direct. Lew could detect no obvious relationship between the North Atlantic Oscillation and lobster abundance along the coasts of either Maine or Rhode Island. But clearly, something big was going on.

For the study of lobster ecology in the Gulf of Maine in the years to come, the primary challenge would be to determine the trajectories of actual larvae, from the locations where they hatched through the currents of the sea to the locations where they settled to the bottom—and to do so while monitoring the abundance of lobsters at various stages in the animal's life cycle: numbers of eggs, larvae, babies, and adults.

It would be a threefold task. First, the distribution of both nearshore and offshore egg-producing lobsters throughout the gulf would have to be mapped. Diane Cowan's sonar-tracking project, Carl Wilson's counting and tagging of eggers during sea sampling, and Bob Steneck's ROV dives would provide information on where mother lobsters were hatching their eggs and in what numbers.

Second, analysis of ocean movements by Lew Incze and other oceanographers would provide information on where the larvae were going after they hatched, both on short trajectories near shore and long trajectories down the coast.

Third, vacuum sampling in lobster nurseries by Rick Wahle, Bob Steneck, Carl Wilson, and other biologists along the New England coast, as well as tidal sampling by volunteers working through Diane Cowan's Lobster Conservancy, would provide information on where the larvae were settling on the bottom, and in what numbers.

By monitoring all three sets of information simultaneously, ecologists might one day be able to identify the causes—and predict the effects—of fluctuations in lobster abundance as they occurred.

~

Lobster science had come a long way since 1895. Yet in a sense, science had simply confirmed the conclusions that lobstermen like Warren Fernald, his sons, and Jack Merrill had arrived at themselves. After the lobster crash of the 1920s and 1930s, the lobster industry had recognized the need to protect the supply of eggs. Beyond that, most lobstermen believed that fluctuations in the catch were beyond their ability to control. Even at the beginning of the twenty-first century, buoyed by a decade of extraordinary catches, fishermen like Bruce Fernald and Jack Merrill didn't expect the huge hauls to last.

Bruce and Jack believed they had done their part by protecting egg-bearing lobsters, undersized lobsters, and oversize lobsters, and they would reap whatever reward Mother Nature saw fit to bestow. The lobstermen of Little Cranberry Island felt that if catches declined to previous levels, it wouldn't be the result of overfishing, but of the lobster population passing through a natural upswing and entering a natural downswing. The study of lobster ecology had given some scientists a similar view.

The government's assessment, by contrast, had changed little. The National Marine Fisheries Service still listed the American lobster as "overfished," and scientists at the agency still recommended raising the minimum size.

These scientists were actually saying two different things when they used the term "overfished." One was the problem that most people think of when they hear the word—exploitation of a marine population beyond the point of long-term sustainability. But in the parlance of fisheries science, "overfished" can also refer to the short-term problem of animals being harvested before they have grown big enough—a farmer wouldn't cut down Christmas trees, for instance, when they're only a few feet tall. A population can be deemed overfished in this sense even if the prospects for long-term sustainability are good.

Both types of overfishing worry the National Marine Fisheries Service because the agency is charged, on the one hand, with building fish stocks over the long term and, on the other, with encouraging their efficient exploitation in the short term. It may seem a paradoxical pair of goals. But by increasing the minimum size of lobster, government scientists believed, they could end both types of overfishing. More lobsters would make eggs, and more lobsters would be caught after they'd grown larger. The former would help ensure the long-term survival of the population. The latter would increase the total amount of lobster meat available for human consumption, from the same number of animals.

After the acrimony between the lobster industry and government scientists peaked during the 1980s, however, state legislatures had circumvented the authority of these scientists, weakening their ability to impose management measures directly. In the 1990s the state of Maine had initiated an experiment called "comanagement." Government scientists would still determine the overall goals for the fishery, but lobstermen themselves would choose the specific measures that would allow them to reach those goals.

On Little Cranberry Island, Mark Fernald and Bruce Fernald were elected to the regional council of lobstermen responsible for the Mount Desert Island area. The council voted to cap the number of traps each lobsterman could use at eight hundred, and also to limit the number of new lobstermen that could enter the fishery. The result was a confusing mix of unintended consequences. Lobstermen with fewer than eight hundred traps rushed to reach the new limit—they feared losing out if the cap was subsequently lowered. Meanwhile, teenagers bought boats and started setting traps—they feared losing the chance to become lobstermen later. The overall effect was to increase trapping effort, not decrease it.

As lobstermen discovered the pitfalls of managing themselves, they also discovered some benefits. The Maine Lobstermen's Association still opposed raising the minimum size, and the new system of comanagement let the MLA participate directly in the making of policy. When government scientists argued that the lobster population wasn't producing enough eggs, the MLA countered that V-notching and Maine's oversize law ensured sufficient egg production. When government scientists argued that a larger minimum size would benefit lobsters and fishermen alike, the MLA countered that raising the minimum size would, by the government's own calculations, provide only a marginal benefit to egg production while incurring significant risk for lobstermen—consumers might still balk at buying more expensive lobsters.

On the surface little had changed. But instead of fuming from the sidelines, lobstermen were sitting at the table and casting votes. Instead of fighting for access to scientific evidence, lobstermen had their own science to present. When they took the sea-sampling data they'd helped collect to the National Marine Fisheries Service, and offered to make V-notching mandatory instead of voluntary, government scientists were forced to acknowledge that V-notching satisfied most of their stated concerns about egg production over the long term.

Ed Blackmore, now retired from his position as president of the MLA, was pleased. An approach he'd advocated for decades had come to fruition. "If fishermen are part of the problem," Ed had always liked to say, "then fishermen have to be part of the solution."

Other New England fisheries—the troubled cod industry in particular—began to look to Maine's lobster industry as an example. Decades of centralized government oversight of groundfish had resulted in many draconian regulations, but few practical solutions. Cod fishermen wondered whether local stewardship might be a better alternative—for both them and the fish.

Meanwhile, government scientists continued to refer to the lobster population in the Gulf of Maine as "overfished." In the technical parlance of fisheries economics the use of the word was justifiable, since raising the minimum size could still result in a higher total yield of lobster meat. But as a description of the biological sustainability of the lobster population, the continued use of the word "overfished" did not reflect the reality most Maine lobstermen saw in their traps.

Back on Little Cranberry Island, overfishing was the least of the lobstermen's fears. In 1999 disease had destroyed most of the lobsters in Long Island Sound. No one knew for certain what had caused the epidemic, though explanations included parasites, warm water, overcrowding, and the spraying of pesticides, especially for West Nile virus. For Jack Merrill the danger posed by pesticides was particularly worrying. To demonstrate how similar the nervous systems of lobsters were to those of insects, he helped his daughter with a science project at school. While her classmates watched, she sprinkled a few drops of a household pesticide over a tank containing a lobster. Within seconds the animal convulsed and was dead. To

Jack, toxic runoff seemed a far more immediate danger than overfishing.

Barring such threats, on Little Cranberry Island there was at least one fisherman who thought a natural downswing in lobster catches might be just what the industry needed.

"I always relish a shakeout," Warren Fernald would opine with a grin. For him, the discipline imposed by Mother Nature was to be welcomed. "Sometimes scoundrels get into the fishery. After a shakeout they don't do so well. The guys that have been hanging in there do okay."

EPILOGUE

Hauling In, 2001

The *Double Trouble*'s new engine had cost Bruce Fernald twenty thousand dollars, and the whirring monster blasted a well-tuned howl from the exhaust stack as Bruce churned the boat through another circle across the sea. But it was the few hundred bucks he'd spent on a different piece of equipment that had most dramatically improved his efficiency hauling traps. Bruce had finally started wearing glasses. This morning, though, even they didn't seem to help.

"Come here, you son of a bitch!" Bruce shouted.

Jason Pickering, the *Double Trouble*'s sternman, rushed to the helm with a pained expression on his face.

"No, no, not you," Bruce said. "I was talking to the buoy."

When a buoy eluded Bruce his tantrums were legendary. Picking out a Styrofoam bullet on twenty square miles of sea pimpled with whitecaps was an incomparable form of aggravation, even with four satellites telling the GPS unit that the buoy was already aboard the boat.

"Tide," Bruce said, invoking every lobsterman's nemesis, "you turned early on me, didn't you?"

When the tide is running hard, a stationary lobster buoy looks like it is streaking across the surface, a bubbly wake boiling behind it. As long as the buoy doesn't get dragged under, it is relatively simple to locate because it gets tugged in a predictable direction—often that is toward shore if the tide is flood and away from shore if it is ebb, depending on local currents. But during the half hour or so when the tide switches directions, the position of a buoy is about as predictable as that of a helium balloon on a hundred-foot tether in a shifting breeze.

A tide calendar would tell you that the sea floods toward land for six hours and six minutes and then reverses direction and ebbs for the same length of time, this cycle occurring roughly twice a day. During a strong flood or ebb tide, a string of buoys ought to behave like a row of balloons in a steady wind, each tied to a brick, strings pulled taut at an angle. A helicopter pilot, having dropped the bricks single file in line with the wind, could circle back and fly straight into the wind, pick up each balloon, and keep flying while he reeled in the string and approached the brick underneath, and so on for the next balloon and brick. Bruce liked to haul his traps in a similar way. Whenever the terrain of the seafloor allowed it, he set his strings of traps more or less parallel to the flow of the tide. When it came time to haul them a few days later, he'd drive the boat into the oncoming current and haul the row of traps from downstream.

This morning Bruce had planned to arrive at the offshore end of his first string of traps while the tide was still ebbing. Then he would haul toward land against the tide and finish the string just as the tide changed directions. With the tide having turned to flood, he would wheel the boat around and haul the next string as he headed back out to sea. But Bruce was late and the tide was early and he couldn't even locate the damn end buoy.

"Aw, why?" Bruce grumbled, fiddling with the GPS unit. "There's really just no need at all of this unnecessary bullshit."

Gritting his teeth in frustration, Bruce punched the throttle. The new engine roared and the *Double Trouble* surged forward. Bruce would have to run to the other end of the string and start hauling from there.

So far this season, even when Bruce had managed to hit his buoys like bull's-eyes, hauling his traps had hardly been worth the effort. The previous summer the lobsters had come on early and strong and the catches had been huge. In July the government stock assessment had warned of overfishing, but by then so many shedders were filling Bruce's traps that it had been hard to imagine anything could be wrong. Then something strange had happened. The catches had dried up. Long before the season's usual finale in November the lobsters simply disappeared—almost as if the fishermen had caught every last one.

By Christmas Bruce had stacked his traps in the yard of his empty island house and, like a bear settling in to hibernate, curled up in the rental on Mount Desert Island with Barb and the boys. A winter of snowstorms and bitter temperatures had only worsened when Bob Steneck, Rick Wahle, and Lew Incze announced that they were predicting a decline in the lobster population. The frigid wind and blustering snow had dragged into March, but Bruce had motored his boat out to the island anyway, hacked at the ice that had frozen his traps to the ground, and one by one pulled them up and set them back into the sea. Perhaps he should have left them in the yard. The spring catch had been miserable.

Come June the family had packed their belongings into boxes and boated back out to the white-and-blue house on Little Cranberry for their final few months together before the boys left for college. Now it was late July and there was still no sign of shedders in the traps. When Bruce's alarm had gone off at four thirty that morning he'd been in the middle of a nightmare. He'd been walking across a dry plain through a row of dusty shacks, pulling open one door after another only to find that each shack was empty. Bruce slowed the *Double Trouble* and stared at his GPS unit.

"O*kay*!" he yelled, stamping his foot. "Should be right here."

Jason stuffed a couple of bait bags with herring and readied himself. Once a lobster boat started down a string of traps, the sternman didn't rest until it reached the end, even if the traps were bare of lobsters.

"I see ya," Bruce finally muttered, spinning the wheel and pulsing the throttle. A few yards from the buoy Bruce idled the engine and plucked his work gloves from the bulkhead, dunked them in the hot-water barrel, and wrung them out before pulling them on.

"We've got twenty pair here," he told Jason, indicating that the string had two traps on each of the twenty buoys. "And I sure as hell hope they look better than they did last time."

Bruce motored up on the buoy, leaned over the rail, and snagged it with his gaff. He tugged it aboard and flipped the rope through the pulley hanging over the rail, then down through another pulley and into the hydraulic hauler by his waist. He revved the engine, and the rope screamed through the hauler. After thirty years it was no wonder Bruce was nearly deaf in one ear. Rope flailed out the bottom and heaped itself in loops at his feet. The knot that tied the sinking line to the floating line slammed into the hauler and popped the rope out, forcing Bruce to flip it back into the spinning sheaves with his hand, a move that could cost a lobsterman the tip of a finger. With the knot safely through, Bruce crouched like a snake charmer, gathered up the coils, and stuffed them into the hot-water barrel to burn off the sea grass.

When the head trap banged against the hull, Bruce halted the hauler and hefted the trap aboard by its rope bridle. Spinning the trap lengthwise onto the rail, he forced the slack line back into the pulley and revved the hauler again to bring up the tail trap. Jason flicked the elastic clasps free from the head trap, threw open the wire mesh door, and started tossing an entire underwater world back overboard, including sand crabs, green crabs, hermit crabs, little black spider crabs, whelks, starfish, strands of kelp, pebbles, periwinkles, shrimp, and whore's eggs—the lobsterman's appellation for the nettlesome sea urchin. Other items from the briny deep dribbled out through the trap mesh and onto the deck or Jason's overalls—eel-like butterfish, long-legged sea stars, jumping sea fleas, and bodiless sea spiders. Untying the bait bag, Jason flung the spent herring to a flock of gulls hovering off his shoulder. The squawking birds plunged into the water to fight over the morsels of rotten fish.

A few seconds later, the tail trap broke the surface and

Bruce pulled it aboard. Punching the boat into gear with one hand, he reached into the trap with the other, grasped a gasping sculpin by its tail, and whapped the fish against the side of the bait bin before tossing it in. With a knife Jason sawed into the sculpin from behind its right fin through its skull to its jaw. He hung the jiggling fish by the incision onto the string of a bag full of bait, the sculpin's dripping entrails to serve as an additional enticement.

By the time Jason was locking down the door of the head trap, Bruce was already racing the boat toward the next buoy. Jason knotted a bait bag into the tail trap and closed it just as Bruce signaled him to throw. Jason shoved the head trap overboard and leaped backward. Rope flew out of the hot-water barrel, spraying steaming globs of seaweed across the cabin ceiling. While Bruce threw the tail trap, Jason sidestepped the loops of rope flailing across the deck and leaned into the bait bin to stuff another pair of bags. The next head trap banged against the hull just as Jason was tightening the drawstrings. He dunked his hands in the hot-water barrel to rub the fish guts off his gloves before turning to open the next two traps.

They too were full of everything but lobsters.

~

In the waters off Little Cranberry Island, Bob Steneck and Carl Wilson had just completed another set of dives from the R/V *Connecticut,* sending down the *Phantom* to look for large lobsters. The scientists had spoken with Bruce Fernald and Jack Merrill that day, and the concerns expressed by the fishermen echoed what Bob and Carl had heard from lobstermen in other parts of the coast. Catches that spring had been slow everywhere, and now it was the end of July and the shedders had yet to appear. Lobstermen were worried about the prospect of a decline. They were also worried that if a decline did occur, the government would blame them for it.

As the *Connecticut* steamed away from Little Cranberry Island, Bob and Carl counted up the number of lobsters the robot had spotted on the bottom. So far the lobstermen's worries seemed premature. That afternoon Bob had mentioned to Bruce and Jack that he'd started to see shedders hiding among the rocks, and a review of the day's data confirmed that impression. If a decline was coming, Bob and Carl didn't think it had arrived yet. Nor did they think that lobstermen should be blamed.

In a classic overfishing scenario like the one that had occurred in the cod fishery, fishermen depleted the resource by ratcheting up their effort on a population that was already diminishing. The situation in the lobster fishery was quite different. Lobstermen hadn't been depleting the resource, but rather ratcheting up their effort in order to take advantage of a burst of excess supply.

Not even the ecologists could say precisely what had caused this increase in the supply of lobsters. Some combination of ecological processes had allowed more lobsters to survive the transition from larva to adult. Perhaps a fall in fish predation had primed the pump, and a subsequent shift in ocean conditions had changed larval delivery patterns and water temperatures, thus opening up new nursery grounds and widening the demographic bottleneck. Even the subject Bob had originally come to Maine to study—sea urchins eating kelp—was probably relevant. The recent boom in harvesting urchins for the sushi market had allowed kelp to flourish, which in turn had created more hiding places for little lobsters.

Regardless of what had caused the lobster population to expand, the response of fishermen had been straightforward. They had set more traps, built bigger boats, worked harder, and added men and women to their ranks. In fact, Bob and Carl were concerned less that lobstermen were overfishing the lobsters than the reverse—that the lobsters were, in a sense, overfishing the lobstermen.

This way of looking at the situation shed new light on the issue of overfishing. Carl had recently become the state of Maine's chief lobster biologist—he took the position after Diane Cowan stepped down—and he faced a problem that wasn't so much biological as economic. Even if lobster catches declined by 60 percent, setting off every alarm bell in New England, the fishery would be returning to the level of catches that Warren Fernald's generation had hauled up for half a century. The population that would suffer might not be the lobsters but the fishermen, especially those who had invested heavily in equipment and grown accustomed to a high standard of living. Warren Fernald might have called them scoundrels, but their ranks would include any lobsterman who hadn't saved for a rainy day.

Ironically, the prospect of a drop in catches also shed new light on the conservation strategy that government scientists had been recommending for nearly twenty years—raising the minimum size. Lobstermen had shown V-notching to be so effective at ensuring egg production that an increase in the minimum size seemed to offer little additional benefit. Yet it was possible that the lobster industry's reliance on V-notched lobsters had left lobstermen vulnerable to short-term declines caused by oceanographic conditions. Increasing the minimum size might reduce this volatility.

The reason for this had to do with the general tendency of large lobsters to inhabit deeper water, farther from shore, while small lobsters inhabited shallow water, closer to the coast. All brooding females were capable of moving into a variety of depths, but the protection afforded by V-notching—along with Maine's oversize law—allowed individual females to grow ever larger. Overall, they were more likely to inhabit deeper water. They hatched a great many eggs, and their larvae could be carried by powerful offshore currents to fill the coastal nurseries. By the same token, however, all their larvae could be lost if those currents shifted away from the coast. By contrast, smaller female lobsters were likely to hatch their eggs in shallower water closer to shore, and their larvae were more likely to be retained in local currents. At least some of their larvae had a good chance of making it to the nurseries every year.

An increase in the minimum legal size was likely to beef up the ranks of these smaller females. The additional contribution to egg production overall might be small, but swings in lobster settlement might become less severe, generating steadier catches from one year to the next.

Ultimately, though, declines caused by the vagaries of ocean currents or the shifting forces of climatic oscillations would be impossible to prevent. With Rick Wahle's ongoing settlement index, Lew Incze's GoMOOS ocean-observation system, and Bob Steneck's scuba surveys, Carl believed that science had a chance to warn lobstermen of impending declines with a new degree of accuracy. If the coming few years proved that to be the case, the lobstermen of Maine would have powerful new tools at their disposal for peering into the future.

With those tools would come grave responsibilities. If lobstermen persisted in trying to extract huge hauls from a population that was shrinking, they really would be overfishing in the classic sense. The population of lobstermen was now larger than before, and better equipped. That would make the lobster stock more vulnerable during a downswing. Carl worried that unless the industry prepared an emergency plan in advance, predictions of a decline might come too late. Recently Rhode Island's lobster catch had plummeted. The state and its fishermen could soon face the impossible task of deciding which seven of every ten lobstermen would give up their boats and traps to save the others. Decisions like that could tear even a tight-knit community like Little Cranberry apart.

With the *Phantom* stowed away for the night, the *Connecticut*'s captain put the indigo hills of Mount Desert Island off his stern and set a course deeper into Down East Maine, where Bob would continue the robot dives. Carl climbed up to the ship's bridge and stepped onto the catwalk to watch the sunset. He had written his graduate thesis on the effect of water temperature on superlobster settlement, and his curiosity burned. In five years of scuba diving in the cobblestone coves off Little Cranberry Island's southern shore, Bob and Carl had still never found babies in any appreciable numbers. Yet last year, Bruce, Jack, and their fellow fishermen at Little Cranberry had hauled in record catches. Where were those lobsters coming from?

A few weeks after the *Connecticut* returned to port, Carl packed scuba gear and an underwater vacuum cleaner and headed Down East again. He'd been examining satellite images of sea-surface temperatures with Lew Incze. On the satellite pictures, the water off the western half of Maine was a warm orange color, while off the eastern half it was a cold green. Presumably that explained why the western cove of Damariscove Island was full of baby lobsters while the similar cove off Little Cranberry Island was all but barren.

On the satellite maps Carl and Lew had noticed a few patches of warm water Down East. Most of them were inland coves located miles inside bays, far removed from prevailing currents. They were the last place Carl would have expected to find a lobster nursery. The Maine coast's most productive nursery—Damariscove Island—was exemplary because it stuck offshore like a catcher's mitt. The warm spots Down East were just the opposite.

Nevertheless, the presence of warm water intrigued him, so Carl decided to have a look. Far inside a Down East bay, he splashed overboard and swam to the shallow bottom. He found patches of cobble, mussels, and kelp—decent enough hiding places for babies. He vacuumed a quadrat, returned to the boat, and dumped out the contents of the mesh bag. It was crawling with little lobsters.

For all the mysteries that lobster scientists had unraveled, more secrets waited to be discovered.

~

The lobstermen of Little Cranberry Island wouldn't need the assistance of submersible robots to find their lobsters. In August the shedders struck. As if to make up for their tardiness, the lobsters swarmed into traps as never before. The holding tanks aboard the *Double Trouble* overflowed, and Bruce took hundred-gallon barrels to sea to contain his extra catch. Even Jack's oversize boat, the *Bottom Dollar,* motored into the harbor every afternoon listing from the weight of the lobsters it carried. For Bruce's sternman, Jason, five lean months were rewarded with weekly paychecks beyond his dreams—even with the price of lobster falling in response to the spike in supply.

The routine aboard the *Double Trouble* was manic. Along with his other duties, Jason had to contend with lobsters piling up faster than he could process them. With the rope screaming through the hauler, Jason would hurry to the bait bin and mash fish parts into bags, then rush to the culling box to measure lobsters with his brass ruler, chucking the shorts overboard and banding the keepers like an assembly-line worker desperate to meet his quota. Lobstermen had long ago done away with wooden plugs for restraining a lobster's claws; now the sternman flicked a rubber band over each claw with a pliers-like banding tool. Jason's wrist was soon aching with the repetition, but he hardly cared. At age twenty he knew what it felt like to strike gold. Bruce would have the next trap on the rail before Jason had reached the bottom of the culling box, and it was back to tugging armloads of glistening shedders from the wire traps and piling them in the box. After that, lobsters flew into any spare bucket, milk crate, or bait tray Jason could find aboard the boat.

"That's it for these," Bruce would finally shout, signaling the end of the string of traps. Bruce would spin the wheel and nail the throttle, and the *Double Trouble* would buck across the waves toward the next end buoy. Jason would have ten minutes to accomplish a series of tasks. He would measure and band the accumulated lobsters strewn around him, then transfer them to the holding tanks and barrels of circulating seawater. He would drain the foul-smelling juice off another hundred-pound tray of bait, shovel herring parts into the bait bin while the speeding boat thrashed against the surf, and finally hose the bait slime, sculpin blood, snails, sea fleas, seaweed, and mud out the scuppers in the stern. Bruce would already have throttled down, wrung the hot water from his gloves, and reached for the gaff to snag a new end buoy.

The *Double Trouble* missed several lucrative fishing days at the end of August when Bruce and Barb drove their twin boys to college, one to Massachusetts, the other to Maryland. For the first time in eighteen years Bruce and Barb would have the island house to themselves. Two days after they set foot back on Little Cranberry, a southerly wind blew in a bank of fog and pressed it up against the coast like a blanket.

Through the fog came poking the bow of a sailboat, the word *Physalia* inscribed on her hull. *Physalia* is the genus name for the stinging jellyfish more commonly known as the Portuguese man-of-war. The man-of-war has atop its body a sail-like crest. Wind, the jellyfish's sole means of locomotion, propells it to unplanned destinations.

The *Physalia* belonged to Bob and Joanne Steneck, on a vacation sail that had brought them to Little Cranberry Island. They moored the boat off the restaurant wharf and rowed ashore. As they had fifteen years earlier, they strolled across the island to its south shore and wandered the cobblestone beach. That evening at the restaurant wharf they met up with Jack and Erica Merrill, Bruce and Barb Fernald, Dan and Katy Fernald, and several other lobstermen and their wives for dinner. The other Fernald brother who was a lobsterman, Mark, couldn't make it. Mark had gotten remarried, and was so busy with his young children that dinner was out of the question.

In late afternoon the wind had shifted from southerly to the trademark southwest breeze of summer, and the fog had lifted to a low, undulating ceiling of wavy clouds, as if the humans were gazing up at the surface of a rippled sea from underwater. Near the *Physalia* the lobster boats rested on their moorings—*Double Trouble, Wind Song, Bottom Dollar*. The lobstermen were exhausted after a day of hauling, but the catches had been huge. They ordered a round of beers. Bruce ordered a steak. Bob ordered a lobster.

The talk turned to the season's strange turn of events, the slow spring and the sudden late-summer spurt of shedders. Halfway into his beer, Bruce turned to Bob.

"Sometimes I really wonder," Bruce said, "why they do that. I mean, why do we catch nothing all spring and all summer, and then just, boom! Suddenly we're up to our armpits in lobsters. Last year the shedders came on wicked early, and we were busy all summer. This summer, nothing till August, then this explosion."

"My guess as to what's going on," Bob said, sucking the meat from one of the legs of his boiled lobster, "is that you've got water temperature playing with the molt cycle."

Bob worked on cracking open the cutter claw as he continued.

"You guys know, of course, that the timing of the shed is temperature driven," he said. "When the water warms up, the lobsters begin to molt. At the minimum legal size, they're molting an average of once or twice a year. But I think warm or cold water can make the difference between one molt or two, and of course it affects the time of year molting occurs."

"Sure, but why was this year so different from last year?" Bruce asked.

"Well," Bob said, dipping a chunk of lobster meat in melted butter, "a colder-than-average winter creates cold water temperatures, right? And a warmer-than-average winter creates warm water."

Bruce nodded. Last winter had been long and cold. The one before that had been mild.

"So that could explain the timing of the catches," Bob went on, wiping his beard and cracking open the crusher claw. "After the warm winter two years ago, your first round of shedders molted up to legal size pretty early—in July. Then what you had was probably a second burst of shedders molting up to size in late August and September. The lobsters dried up because with the water so warm, they had all shed early. By the time November rolled around you'd already caught all your shedders for the year."

"And then we had this harsh winter," Bruce jumped in, "so the spring was slow—because the water was so cold. The water didn't warm up enough for a shed until halfway through the summer."

"That's right," Bob responded, breaking off the lobster's tail and compressing it to ease out the meat. "The main point of what I'm saying is this. Just because the lobsters disappear early, or arrive late, doesn't mean we have a crisis in the fishery. Yes, water temperature can affect settlement too—enough years of cold water might reduce the number of lobsters on the bottom over time. But that's a different phenomenon from the seasonal effects of water temperature on molt cycles, which is what we've been seeing over the past couple of years."

"Yeah," Jack broke in, "but you're still predicting that we're going to get a long-term decline as well, right?"

"We are predicting a decline in some areas, possibly starting next year," Bob answered, dipping the tail meat in butter. "If it occurs, it will be due to reduced settlement. The effects are going to vary regionally. Honestly, I don't know how it's going to affect you guys here."

"I don't think we're going to see a decline yet," Jack said. "I see lots of little lobsters in my traps."

Bruce nodded in agreement. "I've seen more eggers and V-notched lobsters in my traps this year than ever before," he said. "It slows you down, there's so many. I'm not worried about a decline. I'm worried about us having too many lobsters, and the price going all to hell."

"It's not impossible," Bob acknowledged. Then he changed the subject. "So, Bruce, I understand you were in a beer commercial." Bob raised his glass with a grin. "I've been hearing about it for years, but I've never seen it."

Barb rolled her eyes. After fifteen years of sobriety and AA meetings, her husband's claim to fame had become an embarrassment. Bruce blushed, then raised his palms in self-defense.

"Hey," he said, "it was the easiest money I've ever made in my life."

"But there's more," Barb interjected. She described how ABC News had used the clip with Bruce in it as an example of why beer commercials ought to be banned from television, because they were a bad influence.

"No one ever told me *that,*" Bob said, laughing.

Barb sat back, folded her arms, and gave Bruce a dirty look, but there was a smirk on her lips.

"What are the chances," Bob asked, "that I could, you know, see the commercial?"

"What, you want to watch it?" Bruce asked.

"Absolutely."

After dinner the others in the group said their good nights, and Bob and Joanne crammed themselves with Bruce and Barb into the cab of Bruce's fishy pickup and rumbled up the road to the house. Bruce pulled out a videocassette and stuck it in the VCR, and they sat down and watched the Old Milwaukee commercial.

"Oh, Bruce, this is fantastic," Bob exclaimed. "Look at you, you're so young and handsome!"

As much as she disapproved, Barb squeezed Bruce's hand. She had to agree. There was Bruce at the end, nodding in agreement while one of the other lobstermen said, "Boys, it doesn't get any better than this."

As Bruce and Barb sat on the couch in their cozy living room, their children off to college, it seemed that the man on the screen might be right. Fishermen could do worse than protecting their young until it was time to release them into the currents. The rest was up to the sea.

APPENDIX

How to Cook a Lobster

Even people who love eating lobster sometimes feel squeamish when it comes to popping a live lobster in the pot. The fact that lobsters have complex and interesting lives is enough to make even an enthusiastic gourmet ask the obvious question: Is boiling lobsters inhumane?

Every few years, someone makes headlines by rescuing lobsters intended for the pot and releasing them back to the sea. The most famous case occurred in 1994 and drew nationwide attention. Mary Tyler Moore, the actress, fell head over heels for a buff sixty-five-year-old named Spike—a twelve-pound lobster confined to a restaurant tank in Malibu, California. Moore hatched a plan to fly Spike to New England and free him, but the restaurant owner refused to sell the lobster, even for a thousand dollars. When radio commentator Rush Limbaugh learned that Moore had been thwarted, he doubled the ante for Spike to two thousand dollars and offered to eat him. The owner refused to budge, opting to keep Spike as a pet. Limbaugh probably didn't realize that even a Maine lobsterman would have returned a twelve-pound lobster to the sea. In Maine, big males are allowed to wander free so they can mate with egg-producing females. Besides, as every lobsterman knows, lobsters that big don't taste as good as smaller ones.

The following summer Moore took her case to Maine. She teamed up with the animal-rights organization People for the Ethical Treatment of Animals (PETA) to protest the Maine

Lobster Festival, an annual event held in the midcoast town of Rockland. PETA designed a campaign logo with the slogan "Being Boiled Hurts!" Beneath an illustration of a lobster were the words "Lobster Liberation." PETA wasn't fazed by the fact that the lobster in its logo was bright red and thus already dead. PETA, it appeared, had boiled its own mascot.

PETA's cause—and indeed, the very notion of animal welfare—derives historically from the idea that the natural world deserves respect. As it happens, the boiling of live lobsters derives from the same idea. Nearly a century ago, well-heeled rusticators from Boston, New York, and Philadelphia were attracted to Maine's pristine beauty and rugged coastline. Until then, most city dwellers had eaten lobster only from cans. For the elite urbanites privileged enough to vacation in Maine, a live lobster purchased straight from the wharf and boiled in the kitchen represented a kind of communion with the basic elements of nature.

In those days, coastal Mainers themselves considered lobster a meal you ate when you could find nothing better, and inland Mainers seldom even saw a lobster. But by the mid-twentieth century, summer rusticators and tourists had popularized boiled lobster throughout much of the United States. Lobster became a high-class cuisine and a symbol of Maine culture. When Maine's legislature considered designs for a new license plate in the mid-1980s, the winner was a boiled red lobster on a white background.

The new license plate was controversial. Some observers wondered whether a dead animal was an appropriate icon. But most of the objections had nothing to do with animal welfare. Maine has the highest poverty rate in New England. Unable to afford lobster, many Mainers considered it a symbol not of Maine but of wealthy outsiders.

"If you wanted to show typical Maine food," one Maine author said in response to the license plate, "you'd be more accurate with the potato. Or better still, how about macaroni and cheese?" Some car owners used Wite-Out to cover up the crimson crustacean.

The lobster license plate was discontinued in 1999 and replaced with a perky chickadee. But it wasn't long before the state's fishing industry brought the lobster plate back. The new tag was introduced in 2003 as a specialty plate. The redesigned lobster is more attractive than the old one, although it is still red. A portion of the registration fee goes to promote sustainable harvesting of the resource by funding lobster research.

Of the scientists who conduct lobster research, many boil and eat lobster with enthusiasm. The biologists described in this book—Jelle Atema, Stanley Cobb, Diane Cowan, Lewis Incze, Robert Steneck, Richard Wahle, and Carl Wilson—spend countless hours with lobsters in the laboratory and the wild, sharing in the most intimate secrets of lobster life. Yet all of these scientists savor boiled lobster at the end of the day. Knowing more about lobsters makes them a more interesting meal. The growing body of evidence indicating that Maine's lobsters are being harvested sustainably is all the more reason to feel good about eating them.

Although the nervous system of the American lobster has been well studied, scientists have been unable to determine whether a lobster feels pain when thrust into boiling water. There are arguments both ways. Some researchers point out that because lobsters lack an autonomic nervous system, they may be incapable of going into shock, as a vertebrate would, and may continue to receive sensory input until their nerve endings are physically destroyed. What, precisely, those nerve endings sense is not clear. Lobsters have stress receptors, but they do not have identifiable pain receptors. It is possible that some other sort of nerve provides the lobster with a sense akin to pain, and yet lobsters and many other invertebrates do not act pained at the loss of a limb or the infliction of many types of wounds. When a lobster is dropped into a steaming pot, its movements are standard escape responses that occur in any threatening situation, and do not in themselves indicate that the animal is feeling pain, despite PETA's claims to the contrary.

While the nervous system of a lobster is fairly complex and

may be capable of processing pain, scientists have found little evidence that it is more sophisticated than the nervous system of an ant, a housefly, or a mosquito. To be sure, a cogent argument can be made for vegetarianism, but for PETA to expend effort on "lobster liberation" is perhaps a diversion from the plight of the mammals, fowl, and fish that are slaughtered by the billions for food. For the average home chef, boiling a live lobster may seem harder than frying a hamburger, but most people would agree that the killing of a cow is a more complex moral proposition than the killing of an ant, a housefly, or a mosquito.

What's more, lobster meat is more healthful than hamburger. Many people associate lobster with cholesterol and fat, but that's because lobster has traditionally been served with melted butter. The meat of a lobster is nearly fat free, with twenty times less saturated fat than beef and thirteen times less than skinless chicken breast. Lobster has fewer calories and less cholesterol than beef or chicken. Lobster flesh is packed with beneficial components such as the vitamins A, B_{12}, and E; the minerals calcium, phosphorus, and zinc; and plenty of omega-3 fatty acids, which reduce the risk of heart attack. Lobster can be prepared in numerous ways without butter—for example, chunks of lobster over pasta in a red sauce is a delicious alternative.

The lobster's tomalley, greenish in color and easy to identify, is a combined liver and pancreas. It filters toxins, so lobster meat remains unaffected by shellfish blights like "red tide" and does not transmit diseases the way clams can. Because the tomalley functions as a filter, however, health experts recommend against eating it. Lobster meat is relatively high in sodium, so people on a low-sodium diet should, of course, avoid it.

Unlike the flesh of most animals and fish, lobster meat can develop toxins within several hours of the animal's death. That is why preparation of fresh lobster requires the chef to kill the animal during, or just prior to, cooking. Thus the chef at home is faced with the prospect of dropping the live lobster into a pot of boiling water—unless he or she is stoic enough to employ the method favored by Julia Child and many other gourmet chefs, which is to plunge a knife into the lobster's head.

The standard approach—boiling the lobster live—has given rise to various techniques intended to ease the experience for the lobster, such as heating the water slowly or even hypnotizing the animal by pointing its head down and rubbing its carapace before dropping it into the pot. Researchers in the Department of Animal and Veterinary Science at the University of Maine recently tested these techniques. Slow heating and hypnosis actually prolonged death—in both cases the lobster stayed active for two to three minutes during cooking. By contrast, a lobster placed directly into boiling water without any preparation was active for only sixty to ninety seconds. The one technique that helped shorten the experience for the lobster was chilling the live animal in a freezer for a few minutes until it became dormant—how long depended on the animal's size. When the chilled lobster was placed in a pot with boiling water, half a minute passed before the lobster showed any signs of movement. It was then active for only twenty seconds before all movement ceased.

If the home chef still feels squeamish, another option has become tenable. Experimentation by food-science specialists at the University of Maine has led to improvements in flash-freezing techniques. Whole-cooked frozen lobster can now taste as good as fresh lobster, and many supermarkets carry frozen lobsters alongside live ones.

On Little Cranberry Island, Warren Fernald has been cooking lobsters for more than half a century, and he uses a simple formula.

"I put about an inch of water in the pot," he says. "You don't need to add salt—they've got enough salt in 'em already." When the water boils, he adds the live lobsters. After covering the pot, Warren waits until steam starts to escape from under the lid. "Then I time them—about eighteen minutes to cook hard-shell lobsters, about fifteen minutes for new-shell shedders."

Maybe boiling a lobster live should be considered an opportunity. Just as the urban rusticators of a hundred years ago understood the boiling of lobsters as a kind of communion with nature, so today, killing the animal we eat offers the rare chance to acknowledge the philosophical and perhaps even spiritual dimensions of the web of life that sustains us—all from the safety of our kitchens. Jelle Atema, who in his basement lab in Woods Hole has spent probably the most time of anyone in the world watching the American lobster up close, puts it eloquently.

"While I do not know for certain, I believe that lobsters may feel pain," Jelle says. "When we kill them for food we should do so quickly. But we should also honor them with thoughtful appreciation for what they have done for us. I believe we should strive for this in all corners of our lives."

Author's Note

This book is a work of nonfiction, and all characters, events, and scenes in the book are real. A number of the events I witnessed firsthand, particularly those that involved the lobstermen of Little Cranberry Island after 1995, including episodes of lobster fishing, island life, and scientific research, often aboard boats. More generally, my personal experiences have also contributed to my descriptions of lobster fishing and island living. I have been visiting Little Cranberry Island all my life. I lived on the island year-round from 1996 to 1998 and stayed there for parts of 2001 and 2003. For two complete years, 1996–97, I worked full-time as the sternman aboard Bruce Fernald's lobster boat, the *Double Trouble*. At one point in the text, when I refer to an unnamed sternman on Bruce's boat, I am referring to myself.

For the events that I did not witness, my primary source was interviews. I supplemented the interviews with documentary evidence whenever possible. When documentary evidence was unavailable, I was usually able to corroborate information about an event with at least two sources. All speech that appears in quotation marks was either recorded by me during the event, quoted from a documentary source, or confirmed as an accurate reflection of what was said at the time by all parties to the conversations depicted. When describing scientific research, I studied published reports about the research in peer-reviewed journals to buttress the information gathered during interviews. Readers familiar with the events described in this book may notice that I have taken some minor liberties with chronology. I have done this to simplify the presentation by keeping material organized thematically. The sections within chapters are generally in chronological order.

In a broader sense, I have taken another liberty—I have chosen to tell the story of a particular set of people. As a result, many others who have made major contributions to the Maine lobster fishery are not featured in the text. For example, while the lobstermen of Little Cranberry Island have been active in lobster research, it was the lobstermen of the South Bristol Fishermen's Co-op in midcoast Maine who were pioneers in collaborating with ecologists in academia. Other lobstermen who have been actively involved in science and management but are not mentioned in the text include David Cousens, the current president of the Maine Lobstermen's Association; Leroy Bridges, the former president of the Down East Lobstermen's Association; Walter Day, of Vinalhaven Island; and Brian McLain of New Harbor—to name just a very few. Lobsterman Dick Allen of Rhode Island has been a key player for decades at the interface of science and industry and was involved in some of the events described in this book. Joe Vachon and other members of the Maine Lobstermen's Association played roles as well.

Likewise, although the scientists described in this book have advanced our understanding of the biology, behavior, and ecology of lobsters, they are not the only ones to have done so. Canadian scientists have been studying the American lobster for decades and are too numerous to name, although Alan Campbell, G. P. Ennis, and Douglas Pezzack should be mentioned in addition to those who appear in the text. In the United States, Richard Cooper and Joseph Uzmann of the National Marine Fisheries Service conducted pioneering undersea observations of lobster ecology, using both scuba gear and manned submersibles. Biologists working for Normandeau Associates began sampling lobster larvae off the New Hampshire coast as early as the 1970s. Robert Dow, Jay Krouse, and Michael Fogarty have been influential experts on lobster ecology and population dynamics for many years. Several international conferences—involving about a hundred lobster scientists—set the stage for the research I have focused on here, and dozens of fisheries biologists, economists, and resource managers from state and federal government as well as academia have been involved in lobster-stock assessment and management. In addition, the Island Institute in Rockland, Maine, has contributed in countless ways to lobster research along the Maine coast and has done much to foster collaboration between fishing communities and researchers, especially in Penobscot Bay but also in other places, including Little Cranberry Island.

Several hundred graduate students, undergraduate students, research assistants, and summer interns participated in the scientific research projects described in this book, along with a number of collaborators not mentioned in the text. In Stanley Cobb's laboratory at the University of Rhode Island, the shelter-eviction and annexation experiments were conducted primarily by David O'Neill, while Patricia Rooney built the seawater racetrack for superlobsters and designed the experiments measuring their swimming speeds. In Jelle Atema's lab at the Marine Biological Laboratory in Woods Hole, Stewart Jacobson, Elisa Karnofsky, Susan Oleszko-Szuts, and Lauren Stein assisted Atema with the lobster-mating experiments in the large tanks, and Elisa Karnofsky and Randall Elgin were Atema's collaborators in the nighttime snorkeling studies. Rainer Voigt was involved in much of the work in Atema's lab and assisted Diane Cowan with her observations there. Christa Karavanich was the primary researcher conducting experiments on individual recognition in lobster combat in Atema's lab, and Thomas Breithaupt was the primary researcher studying urine release in lobster combat. Paul Bushmann conducted much of the experimentation on female selection of dominant males in Atema's flume tank, and the dopamine-electrode lobster-backpack experiment was developed and conducted by Jennifer Basil. Independently, Edward Kravitz at Harvard Medical School has led numerous studies on the role of serotonin in lobster aggression. At the University of New Hampshire, the lobster-trap-video experiment was a joint project of Steven Jury, Hunt Howell, Daniel O'Grady, and Winsor Watson. Lewis Incze's circulation modeling and larval-delivery simulation was jointly conducted with Christopher Naimie of the Thayer School of Engineering at Dartmouth College. Neal Pettigrew, Huijie Xue, and Andrew Thomas of the University of Maine are among the oceanographers involved in GoMOOS.

For a journalist to navigate the parallel waterways of a century-old industry on the one hand and a specialized scientific field on the other is a daunting challenge, and no doubt there are many people whose work I have not done justice. I have tried to ensure accuracy throughout this book, and I have received a lot of help in the attempt. The errors that remain are mine.

Further Reading

More than a hundred scientific articles and technical reports from academic journals and government publications served as primary sources for this book, along with dozens of news reports. The following books also served as sources and will be of interest to readers who desire a more in-depth discussion of lobster biology, the culture and history of the Maine coast, and the management of the lobster fishery.

Acheson, James M. *Capturing the Commons: Devising Institutions to Manage the Maine Lobster Industry.* Hanover, NH: University Press of New England, 2003.

———. *The Lobster Gangs of Maine.* Hanover, NH: University Press of New England, 1988.

Brown, Mike. *The Great Lobster Chase: The Real Story of Maine Lobsters and the Men Who Catch Them.* Camden, ME: International Marine Publishing Company, 1985.

Caldwell, Bill. *Islands of Maine: Where America Really Began.* Camden, ME: Down East Books, 1981.

Cobb, J. Stanley, and Bruce F. Phillips. *The Biology and Management of Lobsters.* (2 vols.) New York: Academic Press, 1980.

Conkling, Philip W., and Anne Hayden. *Lobsters Great and Small.* Rockland, ME: Island Institute, 2002.

Dwelley, Hugh L. *A History of Little Cranberry Island, Maine.* Islesford, ME: Islesford Historical Society, 2000.

Factor, Jan Robert, ed. *Biology of the Lobster,* Homarus americanus. San Diego: Academic Press, 1995.

Herrick, Francis H. *The American Lobster: Its Habits and Development.* Washington, DC: U.S. Fish Commission, 1895.

Martin, Kenneth R., and Nathan R. Lipfert. *Lobstering and the Maine Coast.* Bath, ME: Maine Maritime Museum, 1985.

Acknowledgments

Anyone writing today about the Maine lobster industry owes a debt to James Acheson, author of the seminal study *The Lobster Gangs of Maine* (1988) and the more recent *Capturing the Commons: Devising Institutions to Manage the Maine Lobster Industry* (2003). Since the 1970s, Dr. Acheson has applied the methods of anthropological inquiry to a long-term study of lobstering culture on the Maine coast. His decades of research—in the archives and aboard boats—provided a rich context for my own two years of working aboard lobster boats and living cheek by jowl with lobstermen. Dr. Acheson generously shared an early draft of *Capturing the Commons* with me, and I relied heavily on the book for my understanding of the history of Maine's lobster-conservation laws.

In conducting my research I was assisted by capable experts at a number of facilities, including Pamela Shephard-Lupo at the Bigelow Laboratory and Maine Department of Marine Resources Library and Information Center in Boothbay Harbor; the staff of the Ernst Mayr Library of the Museum of Comparative Zoology at Harvard University; and reference librarians at the Massachusetts Historical Society, the Massachusetts State Archives, the Schlesinger Library at Harvard University, and the George H. W. Bush Presidential

Library at Texas A&M University. Robert Bayer of the University of Maine's Lobster Institute took time from his busy schedule to show me a number of research projects, and the Lobster Institute's assistant director, Cathy Billings, patiently fielded a string of follow-up questions and requests. The Island Institute provided generous logistical assistance. The captains, crews, and technicians of the research vessels *Edwin Link, Connecticut,* and *Alice Siegmund* kindly put up with my presence during busy workdays aboard their boats. Ivar Babb at the National Undersea Research Center at the University of Connecticut and Jan Petri at the Harbor Branch Oceanographic Institution provided technical information regarding dive operations.

Among others, the following people shared their knowledge, experience, and resources with me during my work on this book: Gary Allen and Lisa Hall of Great Cranberry Island, Robert Atwan of *Best American Essays,* Bob Bowman of the Center for Coastal Studies, Don Bradford of Downeast Marine Resources, Don Carrigan of WCSH Channel 6, Yong Chen of the School of Marine Sciences at the University of Maine, David Conover of Compass Light Film and Video Production, Christopher Costello of Little Cranberry Island, David Cousens of the Maine Lobstermen's Association, Patrice Farrey of the Maine Lobstermen's Association, Rodney Feldmann of Kent State University, Joseph Fessenden of the Maine Bureau of Marine Patrol, Randy Flood of Downeast Marine Resources, Michael Fogarty of the National Marine Fisheries Service, Joseph Hannibal of the Cleveland Museum of Natural History, Sue Hill of Little Cranberry Island, Josef Idoine of the National Marine Fisheries Service, Joseph Kelley of the School of Marine Sciences at the University of Maine, Christopher Kellogg of the New England Fisheries Management Council, Jay Krouse of the Maine Department of Marine Resources, Fred and Mary Lord of Port Clyde, Steve and Amy Philbrook of Little Cranberry Island, Ted Spurling Sr. of Little Cranberry Island, Heather Stirratt of the Atlantic States Marine Fisheries Commission, Richard Swartz of *Transition* magazine, David Thomas of Little Cranberry Island, Dale Tshudy of Edinboro University of Pennsylvania, James Wilson of the School of Marine Sciences at the University of Maine, and Huijie Xue of the School of Marine Sciences at the University of Maine. Bill Adler of the Massachusetts Lobstermen's Association also provided assistance, as did Paul Urbanus and Skippy Ryan of the Boston Harbor Lobstermen's Co-op. I am indebted to Shannon Fanning for the Irish epigraph that appears on the opening page. I extend my apologies to those not mentioned, and assure them of my appreciation.

The characters who appear in *The Secret Life of Lobsters* are real people leading busy lives. I am grateful to all of them for letting me tell their stories, and for their support during the process of researching the book. Many of them submitted to repeated interviews and endless questioning, sometimes in the comfort of their kitchens and living rooms but often in the middle of a hectic workday aboard a boat, on a wharf, on the beach, or in a lab. The scientists in this book made heroic efforts to help me understand their work. On Little Cranberry Island, Jack Merrill's warmth and thoughtfulness were an inspiration, and Bruce Fernald's generosity and upbeat attitude were an example to follow. Bruce deserves special thanks for employing me as his sternman aboard the *Double Trouble*—stuffing bait bags and hefting traps for him was a privilege. Bruce and Barb Fernald and Paul and Brenda Fernald and their children were the best neighbors I could have asked for during Little Cranberry Island's long winters. The Fernald clan, including Warren and Ann, and many others on Little Cranberry, welcomed me into the community and made me feel at home. *The Secret Life of Lobsters* is, I hope, a way of giving something back.

This project first came to life thanks to the editors of the *Atlantic Monthly*, who in 2001 asked me to write a magazine article on lobster science and fishing. From the beginning, Cullen Murphy guided the project with patience and wisdom, building the foundation for the book. The late Michael Kelly turned his characteristic curiosity to the subject of lobstering. Toby Lester, a first-rate editor, has been a helpful friend as well, beginning with my internship at the magazine and continuing as I contributed articles. Amy Meeker and Yvonne Rolzhausen taught me the value of pursuing accuracy, and many other members of the magazine's staff have provided encouragement, especially Lucie Prinz. I am grateful to William Whitworth for noticing me and bringing me on board in the first place.

My agent, Stuart Krichevsky, shepherded me through the production of this book with perspicacity, good humor, and some valuable editorial suggestions; Shana Cohen contributed her expertise as well as friendship and cheer. My editor at HarperCollins, Hugh Van Dusen, has been hugely enthusiastic from the beginning, spurring me on at all the right moments and providing insightful comments and encouragement throughout the process. The rest of the capable staff at HarperCollins has my appreciation as well.

I owe an unfathomable debt to several people who went beyond the call of duty to help me improve the manuscript. Throughout the two years I was working on the book, Jennifer Hammock of the Woods Hole Oceanographic Institution was a source of brilliant ideas, keenly reasoned advice, and lessons on scientific thinking; in draft after draft she made countless helpful suggestions. Sarah Corson of Southwest Harbor, Maine, who for most of my life has been teaching me how to write, took time from working on her own book to critique my drafts; her observations resulted in dramatic improvements. Michael Vazquez of *Transition* magazine applied his extraordinary editorial mind to the manuscript and made several wide-ranging recommendations that greatly improved the book.

Over the years my family and friends have tolerated my lobster obsession with grace (some have even claimed to enjoy it), and all of them have my gratitude—Sarah Corson, Dick Atlee, Ann Corson, Ashley Corson, Jennifer Hammock, Spencer Boyer, and the Bishop Allen Co-op in particular. I wish that my father, Walter Corson—an environmentalist and a scientist—could have lived to read this book, for it was his quirky passion and unflagging encouragement that made my own peregrinations possible. He has my everlasting thanks.

Although in no way related to the production of this book, the following organizations and agencies, among others, provided the funding and institutional support that made the research described in this book possible in the first place. Without them I would not have had any lobster science to write about: the Boston Foundation; Boston University; the Culpepper Foundation; the Darden Environmental Trust; the Davis Conservation Foundation; the Environmental Protection Agency; the Greater Piscataqua Community Foundation; the Guggenheim Foundation; the Gulf of Maine Regional Research Program; the Humboldt Foundation; the Island Institute; the Kendall Foundation; the Maine Community Foundation; the Maine Department of Marine Resources; the Maine Geological Survey; the Maine Lobster Zone Councils; the Maine Lobstermen's Association; the Maine Outdoor Heritage Fund; the Maine Sea Grant Program; the National Fish and Wildlife Foundation; the National Institutes of Health; the National Marine Fisheries Service; the National Science Foundation; the New Hampshire Sea Grant Program; the NOAA Center for Sponsored Coastal Ocean Research; the NOAA National Environmental Satellite, Data, and Information Service; the NOAA National Sea Grant Program; the NOAA National Undersea Research Program; the Northeast Consortium; the Office of Naval Research; the Pew Foundation for Marine Conservation; the Rhode Island Department of Environmental Management; the Rhode Island Sea Grant Program; the University of Maine Center for Marine Studies; the University of New Hampshire Center for Marine Biology; the UpEast Foundation; the Whitehall Foundation; and the Woods Hole Oceanographic Institution.